Optimization for Robot Modelling with MATLAB

Hazim Nasir Ghafil · Károly Jármai

Optimization for Robot Modelling with MATLAB

Springer

Hazim Nasir Ghafil
Faculty of Mechanical Engineering
and Informatics
University of Miskolc
Miskolc, Hungary

University of Kufa
Najaf, Iraq

Károly Jármai
Faculty of Mechanical Engineering
and Informatics
University of Miskolc
Miskolc, Hungary

ISBN 978-3-030-40409-3   ISBN 978-3-030-40410-9 (eBook)
https://doi.org/10.1007/978-3-030-40410-9

This Springer imprint is published by the registered company Springer Nature Switzerland AG
The registered company address is: Gewerbestrasse 11, 6330 Cham, Switzerland

Preface

Optimization is the issue of finding optimal choices among a set of alternatives, and this definition matches simple questions from daily life to very complex science problems. Robotics is the interdisciplinary science which requires optimal solutions for its problems in many cases. Specifically, robot modelling requires finding the optimal set of joint variables, optimal paths and trajectories, optimal specifications of the actuators, optimal structures, etc. It is worth mentioning that most robot modelling problems can be classified as nondeterministic polynomial time and finding solutions get more complicated as the degrees of freedom of the robot increase. For instance, you can find an inverse kinematic solution of a planar two-degree-of-freedom robot by geometric methods or closed-form solutions, but the situation becomes hopeless when you try to find a closed-form solution for a seven-degree-of-freedom Cartesian robot.

Optimization algorithms work with objective functions, and these objectives can be easily formulated for all types of robots, from simple planar to the very complicated high-degree-of-freedom robot manipulators. In other words, optimization dramatically simplifies the process of finding solutions to robot modelling problems and makes them simple sequential processes. This book introduces robot problems from the perspective of the optimization algorithms in a simple way, avoiding the complexity that is often found in traditional methods. Optimization algorithms are programming techniques; therefore, it is not enough for the reader to just write about equations, formulations, procedures, and so on when we wish to benefit the reader truly. The programming code for each kind of problem should be discussed and written, and that is what we have done throughout this book by providing MATLAB codes.

We hope that after reading this book, the reader can solve robot modelling problems, write the required codes, use different optimization algorithms to solve the same problem, and compare the performance of different optimization techniques on a single problem. In other words, the purpose of this book is to provide an introductory platform for students (and others) to become researchers in optimization and robot modelling fields. Also, the book may assist researchers in these fields and contains many ideas for future works.

Miskolc, Hungary

Hazim Nasir Ghafil
Károly Jármai

Acknowledgements

The described book was carried out as part of the EFOP-3.6.1-16-2016-00011 'Younger and Renewing University—Innovative Knowledge City—institutional development of the University of Miskolc aiming at intelligent specialization' project implemented in the framework of the Széchenyi 2020 Program. The realization of this project is supported by the European Union, co-financed by the European Social Fund.

Authors would like to acknowledge Mrs. Shaymaa Alsamia for preparing Figures of the book. We would like to thank our family members for their patience and support.

About this Book

Chapter one: Manipulators are very important in industry, and employing optimization to solve robot problems is extensively used to simplify the difficulties. Chapter two: Optimization algorithms provide us with solutions to problems that cannot be solved or are hard to be solved by traditional methods. Chapter three: Understanding spatial representation is the base of the robotic discipline and essential for subsequent topics. Chapter four: The kinematics of a robot is the description of the motion of a manipulator without taking consideration of the forces or torques that cause this motion. Within kinematics, time-dependent or time-independent problems like position, orientation, velocity, and acceleration are studied. Chapter five: Velocity kinematics becomes relevant when we speak about industrial robot manipulators because of its importance in manufacturing process. For instance, we have to consider velocity kinematics when we design the controller of manipulators for welding, cutting, painting, grinding, etc. Also, velocity kinematics is essential for subsequent topics like dynamics and stiffness analysis. Chapter six: Developed solutions to the forward kinematics and inverse kinematics are essential for path planning which is a geometric description of the motion and trajectory planning which describes how this motion occurs along the path. Chapter seven: The kinematic equations studied in the previous chapters do not describe the relationship between the motion and its producers, i.e. forces and torques. Thus, it is important to study the effect of the forces and torques on the motion of the robot. Chapter eight: One of the most essential issues in robotics is the optimization of the dimensions of the physical components of the robot and thus mass reduction. This issue can be dealt with by a kinematic and kinetic viewpoint. Chapter nine: Kinematic synthesis is the problem of finding the best robot type for a specific task and best dimensions for that robot type.

Contents

About the Authors

Hazim Nasir Ghafil is Member of the teaching staff at the Mechanical Engineering Department, University of Kufa, Iraq, where he taught MATLAB and numerical analysis for three years before joining the University of Miskolc, Hungary, for his Ph.D. studies. At the University of Miskolc, he is an optimization expert and is currently working on his dissertation on the design of robots from an optimization standpoint at the Faculty of Mechanical Engineering and Informatics. He has published several papers on optimization and robotics.

Dr. Károly Jármai is Professor at the Faculty of Mechanical Engineering at the University of Miskolc, where he graduated as a mechanical engineer and received his doctorate (Dr. univ.) in 1979. He teaches design of steel structures, welded structures, composite structures and optimization in Hungarian and in English for foreign students. His research interests include structural optimization, mathematical programming techniques, and expert systems. He wrote his C.Sc. (Ph.D.) dissertation at the Hungarian Academy of Science in 1988, became European Engineer (Eur. Ing. FEANI, Paris) in 1990, and completed his habilitation (Dr. habil.) at Miskolc in 1995. Having successfully defended his doctor of technical science thesis (D.Sc.) in 1995, he subsequently received awards from the Engineering for Peace Foundation in 1997 and a scholarship as Széchenyi Professor between the years 1997 and 2000. He is Co-author (with József Farkas) of four books in English: *Analysis and Optimum Design of Metal Structures, Economic Design of Metal Structures, Design and Optimization of Metal Structures and Optimum Design of Steel Structures*, and three monographs in Hungarian and has published over 782 professional papers, lecture notes, textbook chapters, and conference papers. He is Founding Member of the International Society for Structural and Multidisciplinary Optimization (ISSMO), Hungarian Delegate, Vice Chairman of commission XV and Subcommittee Chairman XV-F of the International Institute of Welding (IIW). He has held several leading positions at GTE (Hungarian Scientific Society of Mechanical Engineers) and has been President of this society at the University of Miskolc since 1991. He was Visiting Researcher at Chalmers University of Technology in Sweden in 1991 and Visiting Professor at Osaka University in

1996–1997, at the National University of Singapore in 1998, and at the University of Pretoria several times between 2000 and 2005. He was Vice-Rector of the university from 2013 to 2017 for the fields of strategy and research. He currently serves on the editorial boards of several national and international journals.

Chapter 1
Introduction

1.1 Background

The development of mechanisms has passed through different stages since the invention of the wheel; the first form of the mechanism was found in an archaeological excavation in Mesopotamia (modern Iraq), and dates to around 3500 BC (Postgate 2017) (Fig. 1.1). Many kinds of technical literature have been written, and many thoughts have been introduced from that time to develop different types of mechanisms, like Watt and Stephenson mechanisms which are shown in Fig. 1.2. Around the middle of the twentieth-century people began automating mechanisms to do a specific job, taking advantage of new opportunities due to the development of computer science. In the 1960s, these automated machines were unique devices called industrial robots (Craig 2005). Nowadays, robots play a significant role in all aspects of human life (Daneshmand et al. 2017; Spolaôr and Benitti 2017) because of the human tendency to design products with low cost, high quality, and fast production rates. Using human workers, this may be difficult to do; robots are preferred especially in some risky working environments, which should undergo a serious risk assessment (Gopinath and Johansen 2016; Michalos et al. 2014). The vehicle industry and other automotive engineering manufacturers are a perfect area for robotization for the reasons mentioned above.

Axiomatically, robots' specifications depend on their applications which differ from one purpose to another. For instance, there are assembly robots (Kramberger et al. 2017; Makris et al. 2017) which carry heavy parts or PCB manipulators where dynamic loads do not have to be considered—of course, both of the previous examples require precise motion. There are many types of industrial robots, and each is used according to the purpose and desired duty (Spong et al. 2005). The most common type of robots are serial robot manipulators, which are a series of rigid bodies, called links, joined together by means of joints (Ghafil et al. 2015), see Fig. 1.3. In manufacturing lines of the vehicle industry, it is economically undesirable to design all the robot manipulators according to the same criterion because the robot's joints and links are subject to different loads in different lines of production. A robot manipulator in

H. N. Ghafil and K. Jármai (ed.), *Optimization for Robot Modelling with MATLAB*,
https://doi.org/10.1007/978-3-030-40410-9_1

Fig. 1.1 Sumerian wheel: the first form of the mechanism ever found

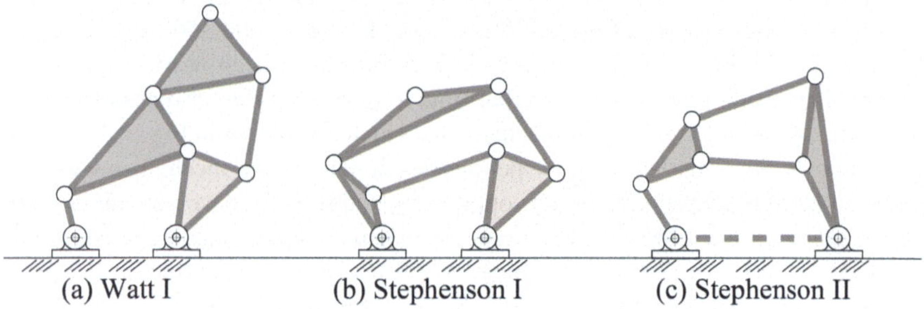

Fig. 1.2 Watt and Stephenson mechanisms

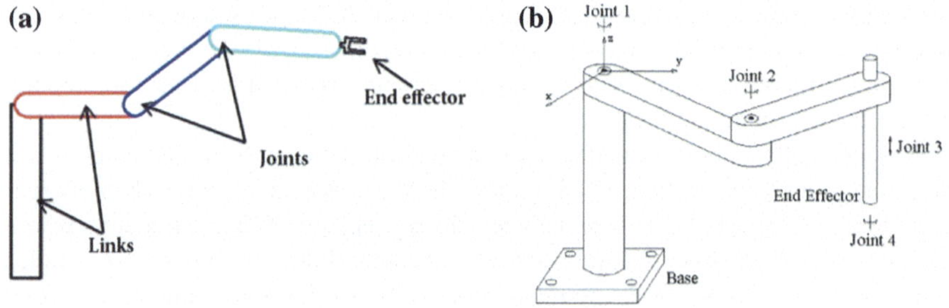

Fig. 1.3 **a** 5R manipulator, **b** RRdR manipulator (SCARA robot)

an assembly line will suffer from more stresses than one in the painting or welding lines. Therefore, there is a need to optimize manipulators links and joints to reach optimum design (Kivelä et al. 2017).

Another fact that should be considered about using robot manipulators in the industry is that the working area or the configuration space of the manipulators may contain either static or dynamic obstacles, which will lead us to supply the robots with path or trajectory planning. These paths might be a predefined set of points (Mathew and Hiremath 2016) in the Cartesian space in the case of a static environment or paths that continuously change due to a dynamic environment (Tuncer and Yildirim 2012). In both cases, these sets of points should be transformed from a configuration space to joints space utilizing of inverse kinematics (Iliukhin et al. 2017).

Industrial robots are spreading rapidly worldwide as a result of the competition among manufacturers in different countries, and reliable statistics refer to the fast growth of robots in the last few years. A world robotics report released by the International Federation of Robotics (IFR) (Heer 2016) shows how the number of industrial robots increased during one year (Figs. 1.4 and 1.5). According to the Robotic Industries Association (Association 2016) in 2016, orders for robots increased by 61% in the field of assembly and 24% in spot welding, while the consumer goods and food industry increased orders for robots by 32%. Figure 1.6 shows how the number of robotics units increased worldwide in the last few years (Ghafil and Jármai 2018). Anyway, many reports have been done to show the worldwide booming of the industrial robots.

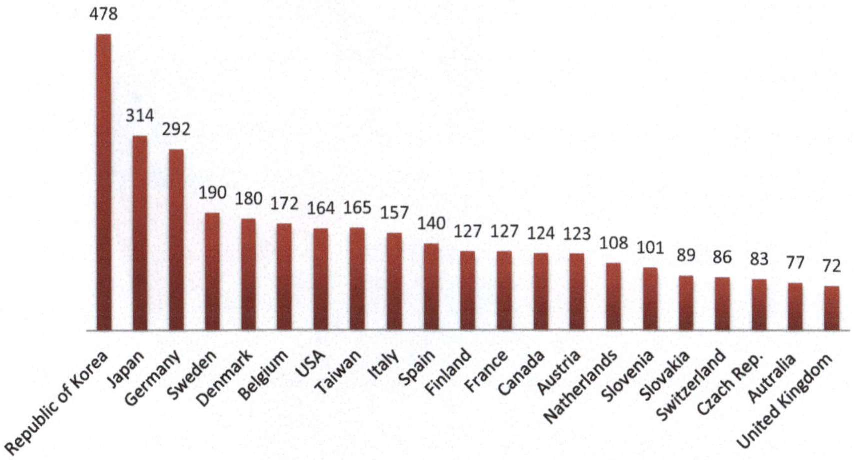

Fig. 1.4 Number of multipurpose industrial robots (all types) per 10,000 employees in the manufacturing industry 2014

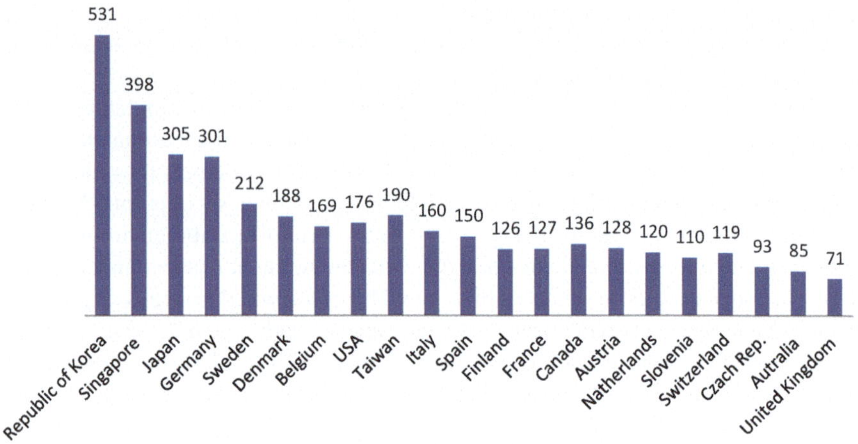

Fig. 1.5 Number of multipurpose industrial robots (all types) per 10,000 employees in the manufacturing industry 2015

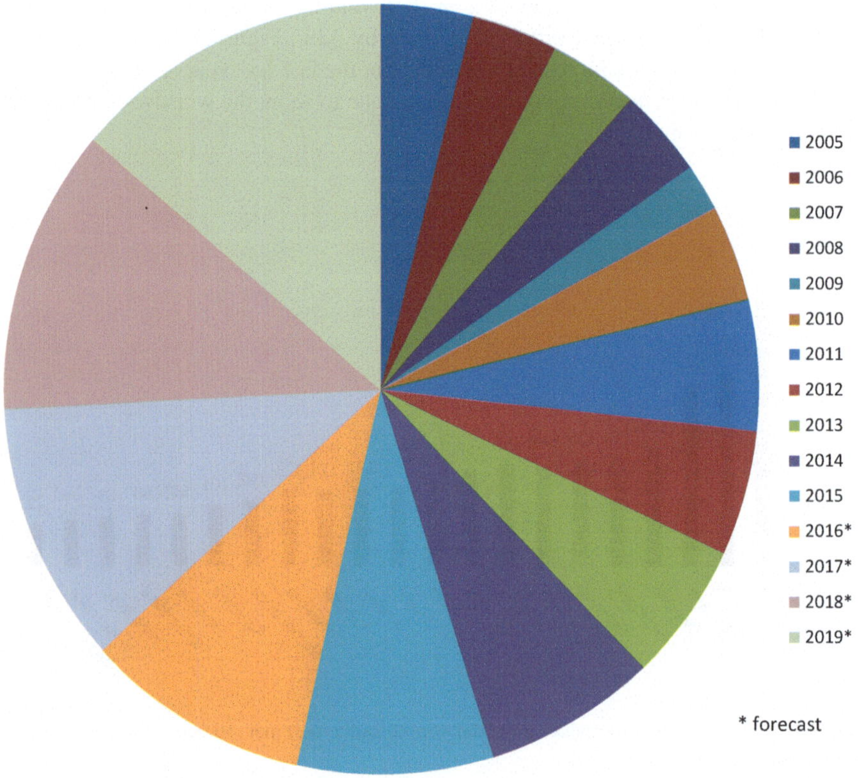

Fig. 1.6 Annual supply for industrial robots worldwide

1.2 Why Optimization?

In many cases where it is hard or impossible to find a theoretical solution to a problem, researchers and scientists tend to use alternative solution methods like numerical analysis or empirical relationships. These alternatives usually give an approximation to the exact solution, and even they do not guarantee a solution for all problems like non-deterministic problems.

For these challenging situations, optimization can be seen as a rescuer, especially when you know that optimization algorithms give satisfactory results and are more accurate than any alternative to the theoretical approaches. For example, let us discuss the following problems in robotics:

1. It is possible to derive a closed-form solution for the inverse kinematics to a planar or ordinary robot manipulators. Still this kind of solution cannot be obtained for robots with seven or more degrees of freedom with complicated motion.
2. Even when a closed-form solution exists, it cannot avoid singularities.
3. Path and trajectory planning and navigation are classified as non-deterministic (NP-hard), so when speaking about the optimal or shortest path, none of the traditional methods are reliable.
4. The same can be said about inverse velocity, inverse acceleration, force control, and mass reduction for the robot arm. All of these problems are hard to find a solution for, and if found, the solution cannot be guaranteed to be the optimum solution.

Consequently, optimization algorithms are the perfect solution for all engineering problems where the minimization or maximization of some quantities is desired. Different algorithms differ in their behaviour: some of them easily fall into the local minima or maxima and others can jump and continue searching for global minima or maxima. We shall discuss that in Chap. 2.

1.3 Application of Optimization in the Industrial Robot Arm

Optimization is a search for local minima and/or maxima at a specific domain in a function (Ghafil 2016), and optimization methods are used in design engineering components by setting a mathematical model for the problem and searching for maximum load or minimum. Many criteria, such as thickness, are put in for problems like reducing aeroplane weight or manufacturing costs (Farkas and Jármai 2013). Researchers have been using some optimization algorithms for decades in many scientific fields (Sai et al. 2016), including the A star, ant colony, bee, genetic, artificial neural network, particle swarm optimization, and harmony search algorithms. Many others, as well as hybrids of two or more algorithms, have been introduced in the optimization process. The purpose of finding the optimal solution is to minimize the consumed time and power, i.e. optimization can be considered an environmentally friendly process as well as it increases the lifetime of the robots, which is economically desirable. In the following, there are examples of applications for industrial robot manipulators that have used optimization techniques to enhance robot performance.

1.3.1 Grinding and Polishing

Robots have been used to finish casted products accurately as well as for removing unwanted edges and polishing the final product. Obviously, if humans perform this work, expert workers are required, and they take more time to do the job compared to robots, which make the process just a regular operation. Grinding robots have been used in different fields, even for grinding submersible structures like dam gates including calculating the material removal rates (Thuot et al. 2013) and for hydroturbine manufacture and repair (Agnard et al. 2015). The material removal rate (Rafieian et al. 2014) is the most important function in this operation, which is preferred to be robotized not just for economic purposes but also for safety, especially in dangerous places like a nuclear power plant, where a grinding robot manipulator has been used for grinding the pipeline system of the plant (Yu 2017). Grinding robots usually have a grinding stone attached to the end-effector, but robotic belt grinding has also been used to increase productivity and to grind more complex surfaces (Zhang et al. 2011). Some studies have focused on the temperature generation and distribution due to the grinding process (Tahvilian et al. 2016), which are important for predicting the properties of crystal structures at a given temperature. The polishing process gives the final fine and shiny appearance to the surfaces, and polishing robots are widely used in different industries like polishing marvel, granite metal sheets, etc. (Dieste et al. 2013). Some studies have used different techniques for the polishing process like machine vision (Yang 2009) and sensor monitoring (Segreto et al. 2015).

1.3.2 Cutting

Manufacturers have been using cutting machines for many years to split objects apart. A variety of complicated shapes can be produced when using CNC or robot manipulators with three axes or more. Cutting robot manipulators have huge applications in different fields of the industry like butchery automation (Liu et al. 2017). They face the challenge of low stiffness (Klimchik et al. 2015) in aerospace or vehicle industries, where cutting forces and gravity result in positioning errors, but some have overcome this problem by introducing an approach to optimize the positioning of the tooltip (Denkena and Lepper 2015). Recent techniques have used damping control system on the cutting tool machine, which has given more refinement to the cutting streams.

1.3.3 Assembly

This is the operation of gathering different parts together to produce a single complicated system like a car engine or the entire car, and it is usually tedious and time-consuming, which makes robot manipulators the right alternative. The classical approach to the robotized assembly operation is graphical motion planning. Though other strategies have been used like machine-vision-based methods (Wan et al. 2017), assembly robots can be equipped with sensors and be well-programmed for safety when working in a cooperative cell with humans (Makris et al. 2017), and collaboration between humans and robots have been used (Wang et al. 2017) because it is convenient for this operation. Among all the processes mentioned above, assembly needs special care because this process includes gathering parts of the vehicle together which may be extremely heavy, i.e. more stresses are applied on the links and joints of the robot manipulators. Therefore, these loads, as well as cyclic load or fatigue, should be considered when the manipulators are designed. Many works have distinguished trajectory planning for assembly process (Valente et al. 2017) and the trade-off between speed and quality.

1.3.4 Painting

Trajectory optimization for painting robots has been done using an offline experimental algorithm by assuming there are no singularities and redundancy (Hyotyniemi 1990). Sometimes painting is not just covering a piece of material with a uniform coating layer; it may take the form of art like Chinese painting. This operation was robotized by decomposing the painting into separate parts and generating a trajectory for each part (Yao and Shao 2005); this is useful especially for the automobile industry where there is sometimes a need to print some logos or patterns on the body.

1.3.5 Welding

The specifications of the welding or strength of the welded area depend on many factors, including the speed of the solder moving along the desired path, i.e. rate of the quantity of the solder has been applied per unit area and the distance between the surface and solder, as well as the shape of the solder movement along the welding path, e.g. zigzag, circular, and so on. These factors make welding by preprogrammed robot manipulators the best choice (Yao and Shao 2005).

Algorithms have been employed for welding robots in several ways. A hybrid algorithm of ant colony and genetic algorithm (Shen 2016) was applied in path planning for a welding robot where the ant colony system gets benefit from the fast convergence of the genetic algorithm, while local optimization has been avoided due to the mutation of the genetic algorithm. Also, the same hybrid algorithm was used in job scheduling for a welding robot (Meng and Chen 2010).

The parameters of welding resistance were optimized to build a relation model that needs the least number of experiments for welding production (Hu et al. 2016) using a combination of orthogonal test with the artificial neural network. Another welding parameter optimization was done by response surface, genetic algorithm, and neural network (Praga-Alejo et al. 2008). Welding robots have been developed to be more handy and productive in different ways such as by using speech recognition (Kumar et al. 2016) to control robot manipulators in a pre-operational set-up welding application. One simulation dynamic software package, RecurDyn, can be used to simulate welding robots and return information like motor specifications or gravity balancing (Wang et al. 2011). A hybrid discrete PSO algorithm was introduced to improve production efficiency of welding robots by improving the path planning of these robots (Wang et al. 2014).

A mobile welding robot with two optimization models for motion planning problem provided a solution for a 'complex all-position welding operation' (Zhang et al. 2013). In some manufacturing cases, when an industrial robot must do multiple tasks, for example, welding different seams, and these tasks occur repeatedly, there is a method (Alatartsev and Ortmeier 2014) to improve the production time of the sequential tasks in production domains in the tasks with freedom of execution. In most industrial robot manipulators, there is the problem of vibration, which may originate from robot motors or be due to the dynamic movements, and the combination of minimum-jerk and minimum-distance trajectories can lead to an effective kinematic scheme (Dong et al. 2015). Two cooperative manipulators in welding process were simulated and investigated theoretically and experimentally (Gan et al. 2014), and the simulation process considered trajectory planning for the two robots with the genetic algorithm adapted to solve this planning.

1.3.6 Optimization and Robot Design

First of all, robot designers have been seeking for an appropriate robot for the proper application to meet the user's economic desires, especially users in the automotive industry, because this industry needs multipurpose robot or sometimes one single robot for many applications. Thus, research began many years ago on using optimization methods to design the appropriate robot for the proper application. Recently, topology optimization has been used to design robots (Briot and Goldsztejn 2018), and algorithms like nonlinear Levenberg–Marquardt (Kivelä et al. 2017) have been used to obtain optimum link length to reduce error in the position and orientation of the end-effector. These examples and many others illustrate how to apply optimization techniques to design the robot itself.

1.3.7 Optimization and Robot Configuration Space

The most critical function in robotics is planning of the motion and path between two known points to find the shortest path, thus consuming both less time and less power for its actuators. The main advantage of planning is that it enables a robot to achieve complex goals (Blackmore and Williams 2006) and move from one configuration to another in a cluttered environment. Path planning is just a geometric operation to describe the path of motion for a robot but does not explain how the motion happens. Path planning for articulated robotic manipulators is usually more challenging than for mobile robots because of the high degrees of freedom. The main key in using a robot manipulator in different industrial applications is that for each application appropriate trajectory planning should occur, where the trajectory describes the motion on a path regarding the velocity, force, and acceleration (Kucuk 2017). In brief, for any industrial robot manipulator application, there is a path consisting of a set of points in space, a 3×1 position vector represents each point, and a 3×3 orientation matrix, the vector, and the orientation matrix were concatenated in a single 4×4 homogeneous transformation matrix. For a trajectory, extra description is needed for the path rather than just a series of single points; as mentioned before, for each point on the path we must specify the velocity, acceleration, and force. Path and trajectory planning occurs in either static or dynamic configuration space, and we shall discuss the mechanism of collision detection later.

1.4 About This Book

This book is about optimization in the robotics field in both the configuration space and the metal structure of the robot arm itself. Different types of heuristics and algorithms will be discussed, described, and built by MATLAB software. This book gives the reader a solid understanding of robot analysis from the viewpoint of optimization. The goals of this book are to provide:

(1) The best knowledge for the reader through many examples and exercises,
(2) The information needed to write a MATLAB code for all the problems of robotics,
(3) Detailed descriptions and to illustrate how to build different types of optimization algorithms from scratch using MATLAB,
(4) Solid knowledge about the optimum design of metal structures of the robot arm,

(5) Simplified methods, especially for inverse problems and avoiding singularities, and
(6) The best knowledge about topological and structural optimization.

For each chapter, there will be many examples and exercises for full benefits of the readership.

1.5 Who Read This Book?

As a prerequisite, readers should have basic knowledge about programming in MATLAB or GNU Octave. Every code will be carefully described, so elementary knowledge of the syntax and function, and the structure of the programming language should be enough. This book is intended for postgraduate and undergraduate students of the third or fourth year. Undergraduate students in the field of mechanical engineering should find it useful also regarding mechanical design. Computer science students will find material relevant to their interests in the field of heuristics and algorithms and can use and modify the presented algorithms for their field of studies, such as image enhancement and data mining.

The reader is not required to read through all chapters of this book because topics can be selected according to the reader's need, for example, one reader can focus on forward kinematics, another can concentrate on inverse kinematics, or yet other can deal with the optimization of the metal structure.

1.6 Outline

We promise a book that can be used as a main document especially for students and researchers who are looking for a guide to developing and programming of optimization problems. This book is intended for people who have different interests. Chapter 2 is intended for all students and researchers in the optimization field, as it explains in detail various types of optimization algorithms; heuristic, metaheuristic, and evolutionary algorithms and some of these algorithms will be written from scratch using MATLAB. Chapter 3 explains the spatial representation of a point and rigid bodies in space, and the reader will be introduced to the concept of homogenous transformations.

Chapter 4 concentrates on the kinematic description of robot manipulators. In this chapter, the reader will learn about the kinematics of the robot and will find many examples of how to build MATLAB code for direct kinematics. There are also code examples for optimization of inverse kinematic problems, using several well-known robots as a case study. At the end, the reader should solve some exercises. Chapter 5 is about velocity and acceleration kinematics. The basic theory is explained, and examples of writing MATLAB codes are given along with some exercises. Chapter 6 presents path and trajectory planning for a robot manipulator. Different algorithms are used as examples and compared to each other to give the reader deeper knowledge of their advantages or disadvantages.

Chapter 7 explains the dynamic analysis of the manipulator and how to find optimum joint forces and torques for specific loading conditions. MATLAB symbolic toolbox is used in this chapter to analyse an example of a robot using Euler–Lagrange equations. Chapter 8 is on structural optimization and stiffness analysis for manipulators. This chapter focuses on finding the stiffness matrix for manipulators, which can then be used for further calculations like vibration control and mechanical design. In this chapter, elastic deformation is only considered for metallic structures. Chapter 9 introduces the concept of kinematic synthesis and explain two automated strategies to find an appropriate robot type with appropriate dimensionality for a specific task.

1.7 Teaching with This Book

This book is useful for courses in robotics, optimization, and mechanical design of multilinkages mechanisms, and the abovementioned topics can be read in this book separately. For optimization course, it is sufficient to teach Chap. 2. For mechanical design course, Chaps. 7, 8, and 9 can be taught. Examples and exercises can be used as tutorials or assignments, and also, algorithms in this book with their codes were discussed in detail. Therefore, students can be asked to modify some of them for specific applications. For graduate students, there are a lot of papers mentioned in this book and that should give them an idea about related works and future work.

References

Agnard S, Liu Z, Hazel B (2015) Material removal and wheel wear models for robotic grinding wheel profiling. Procedia Manufact 2:35–40

Alatartsev S, Ortmeier F (2014) Improving the sequence of robotic tasks with freedom of execution. In: 2014 IEEE/RSJ international conference on intelligent robots and systems, IEEE. Chicago, USA, pp 4503–4510

Association RI (2016) Breaks records for North American robot orders and shipments. https://www.robotics.org/content-detail.cfm/Industrial-Robotics-News/2016-Breaks-Records-for-North-American-Robot-Orders-and-Shipments/content_id/6378. Accessed 14 Nov 2019

Blackmore L, Williams B (2006) Optimal manipulator path planning with obstacles using disjunctive programming. In: american control conference, IEEE. Minneapolis, MN, USA, pp 3200–3202

Briot S, Goldstein A (2018) Topology optimization of industrial robots: application to a five-bar mechanism. Mech Mach Theory 120:30–56

Craig JJ (2005) Introduction to robotics, 3rd edn. Pearson Education International. ISBN 0-13-123629-b

Daneshmand M, Bilici O, Bolotnikova A, Anbarjafari G (2017) Medical robots with potential applications in participatory and opportunistic remote sensing: a review. Robot Auton Syst 95:160–180

Denkena B, Lepper T (2015) Enabling an industrial robot for metal cutting operations. Procedia CIRP 35:79–84

Dieste J, Fernández A, Roba D, Gonzalvo B, Lucas P (2013) Automatic grinding and polishing using spherical robot. Procedia Eng 63:938–946

Dong H, Cong M, Liu D, Wang G (2015) An effective technique to find a robot joint trajectory of minimum global jerk and distance. In: 2015 ieee international conference on information and automation, IEEE, pp 1327–1330

Farkas J, Jármai K (2013) Optimum design of steel structures. Springer Verlag. ISBN 978-3-642-36867-7

Gan Y, Dai X, Da Q (2014) Emulating manual welding process by two cooperative robots. In: Proceedings of the 33rd Chinese control conference, IEEE, pp 8414–8420

Ghafil H (2016) Inverse acceleration solution for robot manipulators using harmony search algorithm. Int J Comput Appl 6(114):1–7

Ghafil H, Mohammed AH, Hadi NH (2015) A virtual reality environment for 5-DOF robot manipulator based on XNA framework. Int J Comput Appl 113(3):33–37

Ghafil HN, Jármai K (2018) Research and application of industrial robot manipulators in vehicle and automotive engineering, a survey. In: 2nd vehicle and automotive engineering conference, lecture notes in mechanical engineering. Springer, pp 611–623

Gopinath V, Johansen K (2016) Risk assessment process for collaborative assembly—a job safety analysis approach. Procedia CIRP 44:199–203

Heer C (2016) World robotics report. https://ifr.org/downloads/press/02_2016/2016FEB_PI__IFR_Roboterdichte_nach_Regionen_QS1.pdf. Accessed 14 Nov 2019

Hu H-Y, Yang J-D, Tian C-L (2016) The research on parameter optimization of power battery pack welding based on neural Network. In: 2016 international conference on robots and intelligent system (ICRIS), IEEE, pp 457–460

Hyotyniemi H (1990) Locally controlled optimization of spray painting robot trajectories. In: Proceedings of the IEEE international workshop on intelligent motion control, IEEE, pp 283–287

Iliukhin V, Mitkovskii K, Bizyanova D, Akopyan A (2017) The modelling of inverse kinematics for 5 DOF manipulator. Procedia Eng 176:498–505

Kivelä T, Mattila J, Puura J (2017) A generic method to optimize a redundant serial robotic manipulator's structure. Autom Constr 81:172–179

Klimchik A, Furet B, Caro S, Pashkevich A (2015) Identification of the manipulator stiffness model parameters in industrial environment. Mech Mach Theory 90:1–22

Kramberger A, Gams A, Nemec B, Chrysostomou D, Madsen O, Ude A (2017) Generalization of orientation trajectories and force-torque profiles for robotic assembly. Robot Auton Syst 98:333–346

Kucuk S (2017) Optimal trajectory generation algorithm for serial and parallel manipulators. Robot Computer-Integr Manufact 48:219–232

Kumar AS, Mallikarjuna K, Krishna AB, Prasad P, Raju M (2016) Parametric studies on motion intensity factors in a Robotic welding using speech recognition. In: 2016 IEEE 6th international conference on advanced computing (IACC), IEEE, pp 415–420

Liu Y, Cong M, Zheng H, Liu D (2017) Porcine automation: Robotic abdomen cutting trajectory planning using machine vision techniques based on global optimization algorithm. Comput Electron Agric 143:193–200

Makris S, Tsarouchi P, Matthaiakis A-S, Athanasatos A, Chatzigeorgiou X, Stefos M, Giavridis K, Aivaliotis S (2017) Dual arm robot in cooperation with humans for flexible assembly. CIRP Ann 66(1):13–16

Mathew R, Hiremath SS (2016) Trajectory tracking and control of differential drive robot for predefined regular geometrical path. Procedia Technol 25:1273–1280

Meng Z, Chen Q (2010) Hybrid genetic-ant colony algorithm based job scheduling method research of arc welding robot. In: The 2010 IEEE international conference on information and automation, IEEE, pp 718–722

Michalos G, Makris S, Spiliotopoulos J, Misios I, Tsarouchi P, Chryssolouris G (2014) ROBO-PARTNER: seamless human-robot cooperation for intelligent, flexible and safe operations in the assembly factories of the future. Procedia CIRP 23:71–76

Postgate N (2017) Early mesopotamia: society and economy at the dawn of history. Taylor & Francis, 392 p. ISBN 1136788638

Praga-Alejo R, Torres-Treviño L, Piña-Monarrez M (2008) Optimization welding process parameters through response surface, neural network and genetic algorithms. In: 2008 electronics, robotics and automotive mechanics conference (CERMA'08), IEEE, pp 393–399

Rafieian F, Girardin F, Liu Z, Thomas M, Hazel B (2014) Angular analysis of the cyclic impacting oscillations in a robotic grinding process. Mech Syst Signal Process 44(1–2):160–176

Sai V-O, Shieh C-S, Lin Y-C, Horng M-F, Nguyen T-T, Le Q-D, Jiang J-Y (2016) Comparative study on recent development of heuristic optimization methods. In: 2016 third international conference on computing measurement control and sensor network (CMCSN), IEEE, pp 68–71

Segreto T, Karam S, Teti R, Ramsing J (2015) Cognitive decision making in multiple sensor monitoring of robot assisted polishing. Procedia Cirp 33:333–338

Shen H (2016) A study of welding robot path planning application based on genetic ant colony hybrid algorithm. In: 2016, IEEE advanced information management, communicates, electronic and automation control conference (IMCEC), IEEE, pp 1743–1746

Spolaôr N, Benitti FBV (2017) Robotics applications grounded in learning theories on tertiary education: a systematic review. Comput Educ 112:97–107

Spong MW, Hutchinson S, Vidyasagar M (2005) Robot modeling and control. Jon Wiley & Sons. Inc. ISBN-100-471-649

Tahvilian AM, Hazel B, Rafieian F, Liu Z, Champliaud H (2016) Force model for impact cutting grinding with a flexible robotic tool holder. Int J Adv Manufact Technol 85(1–4):133–147

Thuot D, Liu Z, Champliaud H, Beaudry J, Richard P-L, Blain M (2013) Remote robotic underwater grinding system and modelling for rectification of hydroelectric structures. Robot Computer-Integr Manufact 29(1):86–95

Tuncer A, Yildirim M (2012) Dynamic path planning of mobile robots with improved genetic algorithm. Comput Electr Eng 38(6):1564–1572

Valente A, Baraldo S, Carpanzano E (2017) Smooth trajectory generation for industrial robots performing high precision assembly processes. CIRP Ann 66(1):17–20

Wan W, Lu F, Wu Z, Harada K (2017) Teaching robots to do object assembly using multi-modal 3D vision. Neurocomputing 259:85–93

Wang X, Li M, Xue L, Ding D, Gu X (2014) Welding robot path optimization based on hybrid discrete PSO. In: 2014 seventh international symposium on computational intelligence and design, IEEE, pp 187–190

Wang XV, Kemény Z, Váncza J, Wang L (2017) Human-robot collaborative assembly in cyber-physical production: classification framework and implementation. CIRP Ann 66(1):5–8

Wang Y-S, Ge L-Z, Xie P-C, Gai Y-X (2011) Dynamic simulation and gravity balancing optimization of spot welding robot based on RecurDyn. In: 2011 IEEE international conference on mechatronics and automation, IEEE, pp 1905–1910

Yang X-S (2009) Firefly algorithms for multimodal optimization. In: International symposium on stochastic algorithms. Springer, pp 169–178

Yao F, Shao G (2005) Painting brush control techniques in Chinese painting robot. In: ROMAN 2005. IEEE international workshop on robot and human interactive communication, IEEE, pp 462–467

Yu P (2017) Research and application of piping inside grinding robots in nuclear power plant. Energy Procedia 127:54–59

Zhang D, Yun C, Song D (2011) Dexterous space optimization for robotic belt grinding. Procedia Eng 15:2762–2766

Zhang T, Chen S, Wu M, Zhao Y, Chen X (2013) Optimal motion planning of all position autonomous mobile welding robot system for fillet seams. IEEE Trans Autom Sci Eng 10(4):1147–1151

Chapter 2
Optimization

2.1 Introduction

Numerous problems in engineering and computer science come with a high level of complexity, and often these problems are nonlinear with complex constraints. Researchers and designers must find a solution to such problems. Finding an optimal solution is similar to searching for a needle in a haystack, if not impossible. Classical optimization (Hillier and Hillier 2003) cannot return a feasible solution to multi-modal or high-order nonlinear problems. Since the 1970s, researchers have been developing nature-inspired algorithms to tackle these kinds of complex problems. These optimization algorithms were introduced after innovation heuristics, and they depend on the description of the problem to find a solution, so they are called meta-heuristics. The objective functions and domain of the variables form a description of the problem. The optimization is called continuous optimization if the variables are continuous in the domain, and is called combinatorial optimization if the variables are discrete in the domain. Some optimization problems need more than one objective functions with multidomain variables; in this case, the problem will be a multiobjective optimization one.

In general, the purpose of optimization is to find the best available value of a function under some criteria. In the simplest explanation, mathematical optimization finds the maximum and minimum of a function. For specific dimensions of a problem, functions may have many local minima and maxima as well as a global minimum and maximum. The best algorithm is one that can search over the domain of the function, exceed local minima or maxima, and reach the global minimum or maximum of the function. On the other hand, algorithms which get stuck in local minima or maxima are not preferable (see Fig. 2.1).

H. N. Ghafil and K. Jármai (ed.), *Optimization for Robot Modelling with MATLAB*,
https://doi.org/10.1007/978-3-030-40410-9_2

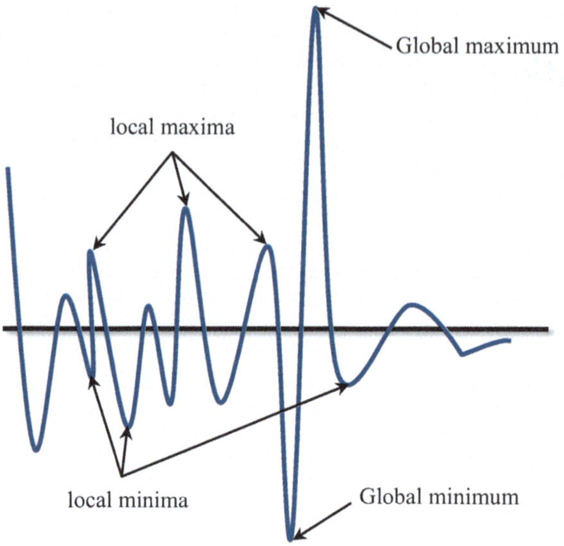

Fig. 2.1 Global and local maximum and minimum

2.2 Ant Algorithm

Consider Fig. 2.2, and assume a barrier is put in the path followed by ants. When ants from both sides to reach the barrier, they have to decide which path they have to take. Initially, there is no information about the new path, so some of the ants go to the left, and some go to the right of the barrier, and while they are moving, they deposit a pheromone. Ants who take the long path will spend more time on it than ants who take the short path, and the evaporation rate of the pheromone is a function of time. Consequently, the concentration of the pheromone on the short path will be greater than on the long path. This is a gradual process happening over time, and for each time cycle, ants follow—probabilistically—the path of more pheromone. Over time, ants abandon the long path due to its weak pheromone traces. For mathematical optimization, ants correspond to candidate solutions, and these solutions behave and develop as the same as real ants. In general, ant algorithms are appropriate for discrete mathematical optimization like travelling salesman problem TSP which will be solved and coded in details in Sect. 6.3. Also, ant algorithms can be used for continuous problems with some modifications to search engine.

2.2.1 Ant System

The ant system (Dorigo 1992), or just AS, can be illustrated as follows: artificial ants make a tour over points in a graph, and this tour is just a group of lines connecting

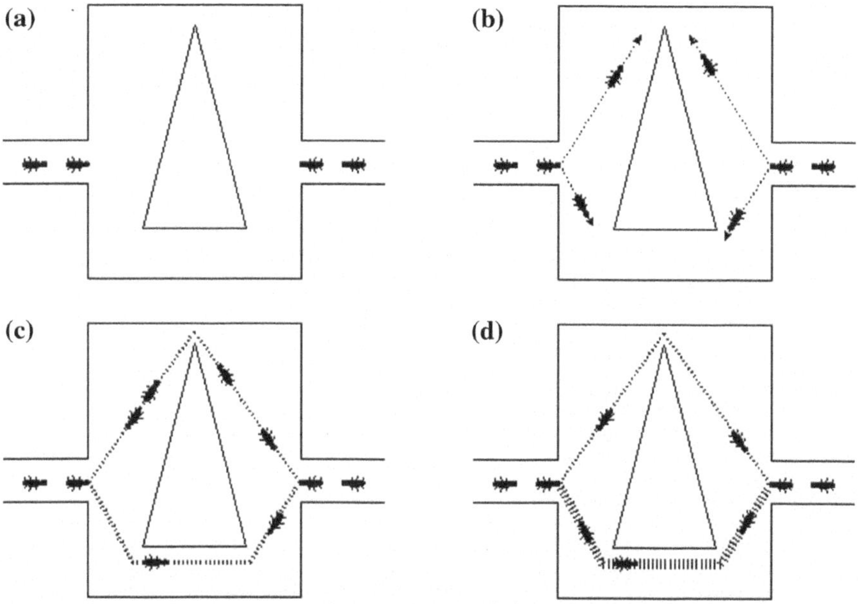

Fig. 2.2 How ants choose their path: **a** ants meet a barrier on their way, **b** randomly some of the ants go to the left or right of the barrier, depositing pheromone as they go, **c** because of evaporation, the long path will have less pheromone than the short path, **d** the pheromone concentration, represented by the dashed line, will become higher on the short path, and this will tempt ants to follow it

the points. Each segment has two quantities: the length of the segment (cost) and the pheromone; the latter is updated continuously during run-time. Ants choose the next point on the graph according to the probabilistic rule called state transition rule:

$$p_k = \left\{ \begin{array}{cc} \frac{[\tau(i,j)][\eta(i,j)]^\beta}{\sum_{u\in J_m(i)}[\tau(i,u)][\eta(i,u)]} & \text{if } (i) \in J_m(i) \\ 0 & \text{otherwise} \end{array} \right\}, \quad (2.1)$$

where τ is the pheromone on the connecting line between points i and j and η is the inverse of the distance between point i and j

$$\eta = \frac{1}{\delta(i, j)}. \quad (2.2)$$

$J_m(i)$ is the set of points that have to be visited by the ant m which is positioned at point i, and u is the set of all points of the tour. β is a weighting parameter to control the importance of the cost versus the pheromone.

When all ants complete their tour among the points, the global update rule is implemented to update the pheromone values using the formula:

$$\tau(i, j) = (1 - \alpha) * \tau(i, j) + \sum_{m=1}^{N} \Delta\tau_m(i, j), \tag{2.3}$$

$$\Delta\tau_m(i, j) = \left\{ \begin{matrix} \frac{1}{L_m} & \text{if } (i, j) \in u \\ 0 & \text{otherwise} \end{matrix} \right\}, \tag{2.4}$$

where $0 < \alpha < 1$ is the pheromone decay parameter and L_m is the length of the path generated by the ant m. N is the total number of artificial ants. The ant system can be summarized as follows:

Algorithm ant system
1- define the number of points for TSP problem
2- Calculate the heuristic information
3- Define the problem parameters
4- Define ant system parameters
5- Initialize heuristic matrix
6- Initialize pheromone matrix
7- for iteration=1 to the maximum number of iterations
8- for ant=1 to the maximum number of ants, each ant chooses a random point to start its tour
9- for decision point = 2 to the maximum number of points of the problem
11- choose the next point, according to equations (2.1) and (2.2)
12- When the tour has built, determine the cost of the tour
13- Update the best tour with the determined minimum cost tour
14- Update pheromone using equations (2.3) and (2.4).
15- stop when the maximum number of iterations is reached.

2.2.2 *Ant Colony System (ACS)*

The ant system provides good results for TSP with variables up to 30, but over that, you will notice that the AS does not give a feasible solution due to the required high number of iterations (it is time-consuming). Ant colony system (ACS) (Dorigo and Gambardella 1997) is one of the improved ant algorithms, and it returns feasible solutions even with a high number of variables. ACS differs from AS in the following:

In the state transition rule, ant m which is positioned at point (i) chooses the next point (j) to build its path according to the following:

$$J = \begin{cases} \arg\max\{[\tau(i, j)][\eta(i, j)]^{\beta}\} & \text{if } q \le q_o \\ \bar{J} & \text{otherwise} \end{cases}. \tag{2.5}$$

Here q is a random number on the period [0, 1], q_o is the weighting parameter $0 < q_o < 1$, and $\bar{J}$ is a random number generated according to Eq. (2.1). The parameter q_o calculates the importance of exploitation of the current best solution versus exploration of a new solution. For every $q \leq q_o$, ants (candidate solutions) use Eq. (2.5) to sample a new point, otherwise they use Eq. (2.1). This state transition rule is called the pseudorandom-proportional rule. During this transition rule, the local pheromone updating rule is applied to the edge (i, j) as follows:

$$\tau(i, j) = (1 - \rho)\,\tau(i, j) + \rho\,\Delta\tau(i, j), \tag{2.6}$$

where ρ is evaporation rate $0 < \rho < 1$ and $\Delta\tau(i, j)$ is calculated using Eq. (2.4). The global updating rule is applied after generating all paths by ants using Eqs. (2.3) and (2.4). ACS is expressed in the following:

Algorithm: ant colony system
1- define the number of points for TSP problem 2- Calculate the heuristic information 3- Define the problem parameters 4- Define ant system parameters 5- Initialize heuristic matrix 6- Initialize pheromone matrix 7- for iteration=1 to the maximum number of iterations 8- for ant=1 to the maximum number of ants, each ant chooses a random point to start its tour 9- for decision point = 2 to the maximum number of points of the problem 11- choose the next point according to equations (26) 12- Apply local pheromone updating rule equation (27) 13- When the tour has built, determine the cost of the tour 14- Update the best tour with the determined minimum cost tour 15- Update global pheromone using equations (24) and (25). 16- stop when the maximum number of iterations is reached.

2.2.3 Ant Colony Optimization for a Continuous Domain

The ant system and ant colony system described above are intended for discrete variables. Combinatorial optimization problems are the only appropriate application for ant algorithms like task scheduling (Srikanth and Geetha 2018), TSP (Colorni et al. 1991, 1992), the quadratic assignment problem (Tsutsui 2007; Whitley et al. 1989), and knapsack problem (Kong and Tian 2006). A modification was made to the ACS to search for optimal solutions with continuous variables on the continuous domain, and this was done by using a probability density function instead of using Eq. (2.1) for more information; the reader can refer to (Socha and Dorigo 2008), and also there is a MATLAB code for this algorithm given in (Heris 2018).

2.3 Flower Pollination Algorithm

The flower pollination algorithm is a biologically inspired algorithm invented by Yang (2012) to simulate the pollination process. The mathematical model of the FPA can be mathematically represented as follows:

1. The pollination process is equivalent to finding a new solution for the optimization problem, and pollen can be considered a solution.
2. In real life, pollination can occur between two flowers on two different plants and called cross-pollination. In mathematical optimization, this process equivalent to searching for a global solution which can be achieved by using Levy flight behaviour, which is explained in many books.
3. Self-pollination and abiotic pollination are searching for a local solution.
4. Flower constancy represents reproduction probability, which is dependent on the similarity among a set of flowers (solution).
5. There is a switch probability $p \in [0, 1]$ to control the interchangeable process between local and global pollination (solution).

Global solution plus flower constancy is represented by

$$X_i^{t+1} = X_i^t + L(X_i^t - g^*), \tag{2.7}$$

where X is the solution of index i within the iteration t, while g^* is the best solution and L is the step size of the pollination process. Levy flight is used to represent the step size, which is defined as follows:

$$\sigma = \left(\frac{\Gamma(1+\lambda)\sin\left(\frac{\pi\lambda}{2}\right)}{\Gamma\left(\frac{1+\lambda}{2}\right)\lambda * 2^{\frac{\lambda-1}{2}}} \right)^{\frac{1}{\lambda}} \tag{2.8}$$

$$u = r\sigma \tag{2.9}$$

$$v = r \tag{2.10}$$

$$\text{step} = \frac{u}{|v|^{\frac{1}{\lambda}}}, \tag{2.11}$$

$$L = 0.01\ \text{step}, \tag{2.12}$$

where Γ is the standard gamma function, λ is a constant taken to be 1.5, r is a random number, and σ, u, v are intermediate variables.

The local solution is represented by

$$X_i^{t+1} = X_i^t + \varepsilon(X_j^t - X_k^t). \tag{2.13}$$

S_j^t and S_k^t are different solutions (pollens) chosen randomly.

Algorithm flower pollination algorithm (Yang 2012)

Objective min or max f(**x**), **x** = (x1, x2, ..., xd)
Initialize a population of n flowers/pollen gametes with random solutions
Find the best solution **g∗** in the initial population
Define a switch probability p ∈ [0, 1]
while (t <MaxGeneration)
for i = 1 : n (all n flowers in the population)
if rand < p,
Draw a (d-dimensional) step vector L which obeys a Levy distribution
Global pollination via equation $X_i^{t+1} = X_i^t + L(X_i^t - g^*)$
else
Draw _ from a uniform distribution in [0,1]
Randomly choose j and k among all the solutions
Do local pollination via $X_i^{t+1} = X_i^t + \varepsilon(X_j^t - X_k^t)$
end if
Evaluate new solutions
If new solutions are better, update them in the population
end for
Find the current best solution **g∗**
end while

2.4 Invasive Weeds Optimization

Weeds are undesirable plants, and this term is associated with cropping systems. Any plant species growing randomly on the farm among crops is defined as a weed. Weeds have faced continuous resistance by herbicides and weeding throughout history, but they still spread and survive. If you observe a certain area with cultivated plants and some weeds, over time, weeds maximize at the expense of the crops; this is really a natural optimization behaviour. The behaviour of weeds motivated the inspiration of invasive weeds optimization (IWO) (Mehrabian and Lucas 2006). The invasive weeds optimization algorithm can be summarized as follows:

1. Initialization

A random number of seeds (solutions) are randomly distributed over the field (search space); this is the first generation.

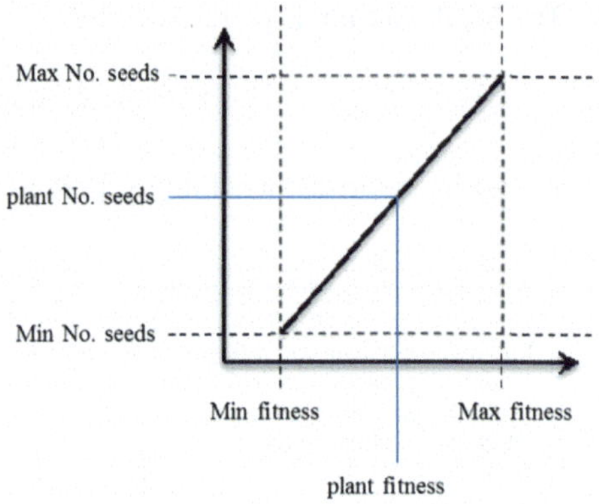

Fig. 2.3 Estimating the number of seeds

2. Reproduction

Seeds are growing into adult plants and generate seeds according to their fitness, this is, the number of generated seeds $\propto$ fitness. In other words, plants produce seeds according to a linear equation representing a line that ends at minimum fitness, or the minimum number of seeds and ends at maximum fitness, or the maximum number of seeds. Figure 2.3 illustrates the mechanism for estimating the number of seeds.

The following formula is valid to calculate the number of seeds:

$$\text{No. of seeds} = \text{floor}\left(\frac{\text{plant fitness}-\text{minimum fitness}}{\text{maximum fitness}-\text{minimum fitness}}\right) \tag{2.14}$$

3. Spatial dispersal

The produced seeds are randomly distributed in the field and grow to the maximum number of plants in the field; these plants are the second generation.

$$\sigma_{\text{iter}} = \frac{(\text{iter}_{\max} - \text{iter})^n}{(\text{iter}_{\max})^n}(\sigma_{\text{initial}} - \sigma_{\text{final}}) + \sigma_{\text{final}}, \tag{2.15}$$

where $\text{iter}_{\max}$ is the maximum number of iterations, σ_{iter} is the standard deviation at the current iteration, and n is the nonlinear modulation index.

4. The process of generating seeds and planting continues to the maximum number of generations.

The algorithm is outlined as follows:

```
Algorithm Invasive weeds optimization
1- Define
Objective Function
Number of Decision Variables
Decision Variables Matrix Size
Lower Bound of Decision Variables
Upper Bound of Decision Variables
2- Define IWO Parameters
Maximum Number of Iterations
Initial Population Size
Maximum Population Size
Minimum Number of Seeds
Maximum Number of Seeds
nonlinear modulation Index
Initial Value of Standard Deviation
Final Value of Standard Deviation
3- initialize the first population
4- for plant = 1 to Maximum Population Size
5- Initialize population with  random positions
6- Evaluate  the positions
7- end
8- for iteration = 1 to Maximum Number of Iterations
9- Update Standard Deviation: equation (2.15)
10- Get Best and Worst Cost Values


11- New plant = 0
12- for plant = 1: Maximum Population Size
13- evaluate plant number of seeds equation (2.14)
14- for offspring = 1: plant number of seeds
15- offspring = population + Standard Deviation *random number
16- Evaluate Offsring
18-  Add Offpsring to the Population: New plant =+ offspring]
19- end
20-  end
21- Population = [plant     New plant]
22-  Sort Population
23- choose the best plants up to the maximum number of population (Delete Extra Members)
24- Store Best Solution Ever Found
 Best plant = Population (1);
end
```

2.5 Bacterial Foraging Optimization

The bacterial foraging optimization algorithm (BFOA) was first described by Liu and Passino (2002) and published in 2002. It is a relatively new swarm intelligence search algorithm. These techniques use the collective intelligence of numerous homogenous individuals. In principle, an individual entity may not be able to solve a problem on its own. However, if numerous individuals form a group, the group's collective intelligence may be enough to solve the task. Bacterial foraging is based on the foraging and the reproduction behaviour of E. coli bacteria colonies. The chemotaxis movement of the group tends towards precious nutrients (global optimum) while avoiding dangerous places (local optima) in the environment. The main parameters of the algorithm are as follows:

Algorithm BFOA Pseudocode of Bacterial Foraging Algorithm (Brownlee 2011)

```
Input: Problemsize   /*Dimension of the search space*/
       Cellsnum      /*Number of cells*/
       Ned           /*The number of elimination-dispersal events*/
       Nre           /*The number of reproduction steps*/
       Nc            /*The number of chemotactic steps*/
       Ns            /*The swimming length*/
       Stepsize      /*The size of step*/
       Ped           /* Elimination-dispersal probability*/
Output: Cellbest
Population ← InitializePopulation(Cellsnum, 1 Problemsize);
for l = 0 to Ned do
  for k = 0 to Nre do
    for j = 0 to Nc do
      ChemotaxisAndSwim(Population, Problemsize, Cellsnum, Ns, Stepsize);
      foreach Cell ∈ Population do
        if Cost(Cell) ≤ Cost(Cellbest) then
          Cellbest ← Cell;
        end
      end
    end
    SortByCellHealth(Population);
    Selected ← SelectByCellHealth(Population, Cellsnum/2);
    Population ← Selected;
    Population ← Selected;
  end
  foreach Cell ∈ Population do
    if Rand() ≤ Ped then
      Cell ← CreateCellAtRandomLocation();
    end
  end
end
return Cellbest;
```

2.6 Bat Algorithm (BATA)

The **bat algorithm (BATA)** is a swarm intelligence metaheuristic method published by Yang (2010) in 2010, based on the echolocation behaviour of bats. Echolocation allows bats to find and identify their prey (global optimum) and avoid obstacles (local optima), even in complete darkness. Echolocation works like sonar, as bats emit a sound pulse and listen for the echo that bounces back from the surrounding environment. During the search, there are changes in the frequency, loudness, and emission rate of the sound of the individual regarding the position of the prey, which informs other bats. The loudness of the emitted sound decreases as the bat approaches the prey, while the rate of pulse emission increases. The bat algorithm combines the advantages of existing swarm intelligence methods, such as particle swarm optimization and the firefly algorithm. The main parameters of the algorithm are as follows:

Algorithm BATA Pseudocode of Bat Algorithm (Brownlee 2011)

```
Input: PopulationSize /*Size of bat population*/
       ProblemSize          /*Dimension of the search space*/
       Loudness             /*Loudness of the sound*/
       PulseRate            /*PulseRate of the sound*/
       Frequency            /*Frequency of the sound*/
 Output: best
Initialize the bat population xi(i = 1,2, ..., population size) and vi
Define pulse Frequency fi at xi
Initialize PulseRates ri and the Loudness Ai
while (t < Max number of iterations)
Generate new solutions by adjusting frequency,
and updating velocities and locations/solutions
   if (rand > ri)
   Select a solution among the best solutions
   Generate a local solution around the selected best solution
   end if
   Generate a new solution by flying randomly
   if (rand < Ai & f(xi) < f(x*))
   Accept the new solutions
   Increase ri and reduce Ai
   end if
Rank the bats and find the current best x*
end while
return best
```

2.7 Bees Algorithm (BA)

Pham et al. (2006) published the bees algorithm (ABC) in 2005. Primarily, it was developed to search for the global optimum of continuous mathematical functions. It belongs to the field of swarm intelligence procedures. The bees algorithm, as the name suggests, was inspired by the foraging behaviour of honey bees. Bee colonies send scout bees to explore the environment and find areas rich in nectar (global or local optima). The scout bees then return to the hive and inform the worker bees about the position and quality of food sources. The number of worker bees sent to each food source depends on these characteristics. The scout bees continually search for good sites, while worker bees accurately explore the already-found spots. Generally, the scout bees are responsible for global search, while worker bees provide better local search. The algorithm is similar to ant colony optimization and particle swarm optimization. However, the rank system of bees makes it unique. The bees algorithm is capable of solving both continuous and combinatorial optimization problems. The main parameters of the algorithm are as follows:

Algorithm BA Pseudocode of Bees Algorithm (Brownlee 2011)

```
Input: Problemsize      /*Dimension of the search space*/
       Beesnum /*Number of bees*/
       Sitesnum /*Number of sites*/
       EliteSitesnum /*Number of elite sites*/
       PatchSizeinit          /*Initial size of patches*/
       EliteBeesnum           /*Number of elite bees*/
       OtherBeesnum           /*Number of other bees*/
Output: Beebest
Population ← InitializePopulation(Beesnum, 1 Problemsize);
while ¬StopCondition() do
  EvaluatePopulation(Population);
  Beebest ← GetBestSolution(Population);
  NextGeneration ← Ø;
  Patchsize ← ( PatchSizeinit × PatchDecreasefactor);
  Sitesbest ← SelectBestSites(Population, Sitesnum);
  foreach Sitei ∈ Sitesbest do
    RecruitedBeesnum ← Ø;
    if i < EliteSitesnum then
      RecruitedBeesnum ← EliteBeesnum;
    else
      RecruitedBeesnum ← OtherBeesnum;
    end
    Neighborhood ← Ø;
    for j to RecruitedBeesnum do
      Neighborhood ← CreateNeighborhoodBee(Sitei, Patchsize);
    end
    NextGeneration ← GetBestSolution(Neighborhood);
  end
  RemainingBeesnum ← (Beesnum- Sitesnum);
  for j to RemainingBeesnum do
    NextGeneration ← CreateRandomBee();
  end
  Population ← NextGeneration;
end
return Beebest;
```

2.8 The Cross-entropy Method

The cross-entropy method (CEM) is a probabilistic optimization algorithm developed by Rubinstein (1997) and published in 1997. The name of the technique comes from Kullback–Leibler cross-entropy divergence as a measure of closeness between two probability distributions. Cross-entropy method is an adaptive importance estimation technique for rare-event probabilities in discrete event simulation systems. Optimization problems can be described as rare-event systems because of the probability of locating an optimal solution by using a random search is a rare-event probability. The method changes the sampling distribution of random search so that the rare event (finding the optimum) is more likely to occur. The main parameters of the algorithm are as follows:

Algorithm CEM Pseudocode of Cross-Entropy Method (Brownlee 2011)

```
Input: Problemsize                /*Dimension of the search space*/
       Samplesnum                 /*Number of samples*/
       UpdateSamplesnum           /*Number of update samples*/
       Learnrate                  /*Learning rate*/
Output: Sbest
Means ← InitializeMeans();Variances ← InitializeVariances(); Sbest ← ∅;
while Max(Variances) ≤ Variancemin do Samples ← 0;
   for i = 0 to Samplesnum do
      Samples ← GenerateSample(Means, Variances);
   end
   EvaluateSamples(Samples); SortSamplesByQuality(Samples);
   if Cost(Samples0) ≤ Cost(Sbest) then Sbest ← Samples0;
   end
   Samplesselected ← SelectBestSamples(Samples, UpdateSamplesnum);
   for i = 0 to Problemsize do
      Meansi ← Meansi + Learnrate × Mean(Samplesselected, i);
      V ariancesi ← V ariancesi + Learnrate ×
      Variance(Samplesselected, i);
   end
end
return Sbest;
```

2.9 Cuckoo Search (CS)

Yang and Deb (2009) developed the cuckoo search (CS) algorithm and published it in 2009. The algorithm was inspired by the brood parasitism of certain cuckoo species, which means they lay their eggs in the nests of other host species. Sometimes if the host birds discover the eggs are not their own, they throw the alien eggs away, or abandon the current nest and build a new one elsewhere. The algorithm randomly distributes a fixed number of nests through the search space. The cuckoos lay their eggs—each cuckoo laying one egg at a time in a randomly chosen nest. If the host bird discovers the egg, it can either throw the egg away or abandon the current nest and build a completely new one elsewhere. Each egg in a nest represents a solution, and a cuckoo egg represents a new solution. The algorithm aims to find a new, potentially better solution (global optimum) to replace the worse (local optima). The best nests with the highest quality of eggs will be members of the next generation. The main parameters of the algorithm are as follows:

Algorithm CS Pseudocode of Cuckoo Search Algorithm (Brownlee 2011)

```
   Input: Problem size          /*Dimension of the search space*/
          NestSize              /*Number of nests*/
          discovery rate        /*Rate of discovery*/
   Output: best
begin
   Generate initial population of NestSize nests xi (i = 1, 2, ..., NestSize)
   while (t < MaxGeneration) or (stop criterion)
      Get a cuckoo randomly by L´evy flights
      evaluate its quality/fitness Fi
      Choose a nest among n (say, j) randomly
      if (Fi > Fj), replace j by the new solution; end
      Worse nests < discovery rate are abandoned, and new ones are built;
      Keep the best solutions (or nests with quality solutions);
      Rank the solutions and find the current best
   end while
end
return best
```

2.10 Cultural Algorithm

The cultural algorithm (CA) was described by Reynolds (1994) and published in 1994. This evolutionary algorithm simulates the cultural evolution of human society. Culture includes the habits, beliefs, and morals of a member of the society. Culture can interact with the environment via positive or negative feedback cycles. As the evolutionary process goes on, individuals gain information about the search space, which is communicated to other individuals in the population. This creates a knowledge base that stores positive feedback about useful areas of the environment (the global optimum), as well as potentially hazardous areas (local optima). This cultural knowledge base is expanded and exploited through generations as situations change. The main parameters of the algorithm are as follows:

Algorithm CA Pseudocode of Cultural Algorithm (Brownlee 2011)

```
   Input: Problemsize            /*Dimension of the search space*/
          Populationnum          /*Size of the polulation*/
   Output: KnowledgeBase
Population ← InitializePopulation(Problemsize, Populationnum);
KnowledgeBase ← InitializeKnowledgebase(Problemsize, Populationnum);
while ¬StopCondition() do
   Evaluate(Population);
   SituationalKnowledgecandidate ← AcceptSituationalKnowledge(Population);
   UpdateSituationalKnowledge(KnowledgeBase, SituationalKnowledgecandidate);
   Children ← ReproduceWithInfluence(Population, KnowledgeBase);
   Population ← Select(Children, Population);
   NormativeKnowledgecandidate ← AcceptNormativeKnowledge(Population);
   UpdateNormativeKnowledge(KnowledgeBase, NormativeKnowledgecandidate);
end
return KnowledgeBase;
```

2.11 Differential Evolution

The differential evolution (DE) algorithm was developed by Storn and Price (1995), published in 1995, and belongs to the field of evolutionary algorithms. Differential evolution is mainly based on Darwin's Theory of Evolution because its main principle is natural selection. The algorithm maintains a population of candidate solutions with recombination, evaluation, and selection as generations unfold. The recombination creates a new candidate solution based on the weighted difference between two randomly selected population members added to a third population member. The main parameters of the algorithm are as follows:

Algorithm DE Pseudocode of Differential Evolution Algorithm (Brownlee 2011)

```
Input: Populationsize          /*Size of the population*/
       Problemsize             /*Dimension of the search space*/
       Weightingfactor         /*Weighting factor*/
       Crossoverrate           /*Crossover rate*/
Output: Sbest
Population ← InitializePopulation(Populationsize, Problemsize);
EvaluatePopulation(Population);
Sbest ← GetBestSolution(Population);
while ¬ StopCondition() do
   NewPopulation ← ∅;
   foreach Pi ∈ Population do
      Si ← NewSample(Pi, Population, Problemsize, Wf, Cr);
      if Cost(Si) ≤ Cost(Pi) then
         NewPopulation ← Si;
      else
         NewPopulation ← Pi;
      end
   end
   Population ← NewPopulation;
   EvaluatePopulation(Population);
   Sbest ← GetBestSolution(Population);
end
return Sbest;
```

2.12 Firefly Algorithm

The firefly (FF) algorithm is a nature-inspired metaheuristic optimization algorithm for multimodal optimization developed by Yang (2009) and published in 2009. The flashing behaviour of fireflies, which is used to communicate with other fireflies, inspired the algorithm. The flashing light is produced by a biological process called bioluminescence and can attract mating partners as well as potential prey. It can also serve as a protective warning mechanism. The rate of flashing and the light intensity are very important characteristics of the communication. The flashing light can be associated with the objective function to be optimized. As a firefly gets closer to a good solution, it will emit more light. The less bright fireflies will move towards the brighter ones. Attractiveness is proportional to the brightness, and it decreases as the distance between fireflies increases. If there is no brighter firefly in its vicinity, then a particular firefly will move randomly. The firefly algorithm has adjustable visibility and is more versatile on attractiveness variations than other techniques, like particle swarm optimization. There are some quite new modifications of the algorithm to make it more efficient (Carbas 2016). The main parameters of the algorithm are as follows:

Algorithm FF Pseudocode of Firefly Algorithm (Brownlee 2011)

```
Input: Problem size          /*Dimension of the search space*/
       PopulationSize        /*Size of the population*/
       γ                     /*Absorption coefficient*/
Output: best
Generate initial population of fireflies xi(i=1,2,...,PopulationSize)
Light intensity Ii at xi is determined by f(xi)
Define light absorption coefficient γ
while (t<MaxGeneration)
   for i=1:n all n fireflies
      for j=1:i all n fireflies
         if (Ij>Ii) Move firefly i towards j in d-dimension;
         end if
         Attractiveness varies with distance r via exp[−γr]
         Evaluate new solutions and update light intensity
      end for j
   end for i
   Rank the fireflies and find the current best
end while
return best
```

2.13 Harmony Search

Harmony Search (HS) was published by Geem et al. (2001) in 2001. The improvisation of jazz musicians inspired the algorithm. When they start a musical performance, they adapt their music to the band, creating musical harmony. If a false sound occurs, each individual of the band makes modifications to improve their performance. The musicians seek harmony through small variations and improvisation, and harmony is taken as a complete candidate solution. The audience's aesthetic appreciation of the harmony represents the cost function. The algorithm has some similarities to the cultural algorithm since the components of the candidate solution are either stochastically created directly from the memory of high-quality solutions, adjusted from the memory of high-quality solutions, or assigned randomly. The main parameters of the algorithm are as follows:

Algorithm HS Pseudocode of Harmony Search Algorithm (Brownlee 2011)

```
Input: Pitchnum               /*Number of pitches*/
       Pitchbounds            /*Pitch bounds*/
       Memorysize             /*Memory size*/
       Consolidationrate      /*Consolidation rate*/
       PitchAdjustrate        /*Pitch adjustrate*/
       Improvisationmax       /*Pitch adjustrate*/
Output: Harmonybest
Harmonies ← InitializeHarmonyMemory(Pitchnum, 1 Pitchbounds, Memorysize);
EvaluateHarmonies(Harmonies);
for i to Improvisationmax do
  Harmony ← ∅;
  foreach Pitchi ∈ Pitchnum do
    if Rand() ≤ Consolidationrate then
      RandomHarmonyipitch ←
      SelectRandomHarmonyPitch(Harmonies, Pitchi);
      if Rand() ≤ PitchAdjustrate then
         Harmonyipitch ← AdjustPitch(RandomHarmonyipitch);
      else
         Harmonyipitch ← RandomHarmonyipitch ;
      end
    else
      Harmonyipitch ← RandomPitch(Pitchbounds);
    end
  end
  EvaluateHarmonies(Harmony);
  if Cost(Harmony) ≤ Cost(Worst(Harmonies)) then
    Worst(Harmonies) ← Harmony;
  end
end
return Harmonybest;
```

2.14 Memetic Algorithm

The memetic algorithm (MA) was developed by Moscato (1989) in 1989. The algorithm simulates the creation and inheritance of cultural information among individuals. A meme is the basic unit of cultural information (an idea, discovery, etc.); this name derives from the biological term gene. Universal Darwinism is the generalization of genes beyond biological-based systems to any system, where discrete units of information can be distributed and subjected to evolutionary changes. The objective of the algorithm is to do a population-based global search, while the individuals explore the promising areas with a local search. The balance between global and local search mechanisms is crucial to ensure the algorithm does not get stuck in a local optimum. A meme is an information about the search, shared between individuals through generations, which influences the evolutionary processes. Memetic algorithm is a duality of genetic and cultural evolution methods. The main parameters of the algorithm are as follows:

Algorithm MA Pseudocode of Memetic Algorithm (Brownlee 2011)

```
Input: ProblemSize          /*Dimension of the search space*/
       Popsize              /*Population size*/
       MemePopsize                  /*Memetic population size*/
Output: Sbest
Population ← InitializePopulation(1 ProblemSize, Popsize);
while ¬StopCondition() do
   foreach Si ∈ Population do
      Sicost ← Cost(Si);
   end
   Sbest ← GetBestSolution(Population);
   Population ← StochasticGlobalSearch(Population);
   MemeticPopulation ← SelectMemeticPopulation(Population, MemePopsize);
   foreach Si ∈ MemeticPopulation do
      Si ← LocalSearch(Si);
   end
end
return Sbest;
```

2.15 Nelder–Mead

The Nelder–Mead (NM) algorithm was named after its creators, as Nelder and Mead (1965) proposed this metaheuristic algorithm, also referred to as the amoeba method, in 1965. Basically, Nelder–Mead is a simplex search algorithm commonly used for nonlinear optimization problems. The algorithm creates random candidate solutions, and each solution has an associated fitness function value. Their fitness orders the candidates, and in each generation, the algorithm attempts to replace the worst solution with a better one. The better solution is chosen from among three candidates: the reflected point, the expanded point, and the contracted point. All of these points lie along a line from the worst point through the centroid. The centroid is in the middle of all points except the worst point. If neither of the points is better than the current worst solution, the amoeba moves all points, except for the best point, halfway towards the best point. The main parameters of the algorithm are as follows:

Algorithm NM Pseudocode of Nelder Mead Algorithm (Brownlee 2011)

```
Input: ProblemSize          /*Dimension of the search space*/
           Amoebasize              /*Amoeba size*/
Output: Sbest
generate amoebaSize random solutions
while not done loop
   compute centroid, reflected
   if reflected is better than best solution then
      compute expanded
      replace worst solution with better of reflected, expanded
   else if reflected is worse than all but worst then
      if reflected is better than worst solution then
         replace worst solution with reflected
      end if
      compute contracted
      if contracted is worse than worst
         shrink the amoeba
      else
         replace worst solution with contracted
      end if
   else
      replace worst solution with reflected
   end if
end loop
return best solution found
```

2.16 Particle Swarm Optimization

The particle swarm optimization (PSO) swarm intelligence metaheuristic algorithm was developed by Kennedy and Eberhart (1995) in 1995. Particle swarm's operation was inspired by the foraging movement of bird and fish swarms. The particles of the swarm move around the search space towards historically good areas. A new position of the particle is influenced by the best-known position in the swarm, also by the best-known position of the whole swarm. The mathematical formulae of the movement to determine the new velocity and position of the particles guide the swarm towards the global optimum. The process is repeated in each generation, while stochastic factors also affect the movement of the particles. There are a great number of papers related to the development and application of PSO (Kennedy 2011; Mortazavi and Toğan 2016). The main parameters of the algorithm are as follows:

Algorithm PSO Pseudocode of Particle Swarm Optimization Algorithm (Brownlee 2011)

```
Input: ProblemSize              /*Dimension of the search space*/
       Population size          /*Size of the population*/
       w                        /*Weighting factor*/
       c1, c2                   /*Learning factors*/
Output: Pg best
Population ← ∅, Pg best ← ∅;
for i = 1 to Population size do
    Pvelocity ← RandomVelocity();
    Pposition ← RandomPosition(Problem size);
    Pcost ← Cost(Pposition);
    Pp best ← Pposition;
    if Pcost ≤ Pg best then Pg best ← Pp best;
    end
end
while ¬StopCondition() do
   foreach P ∈ Population do
      Pvelocity ← UpdateVelocity(Pvelocity, Pg best, Pp best, w, c1, c2);
      Pposition ← UpdatePosition(Pposition, Pvelocity);
      Pcost ← Cost(Pposition);
      if Pcost ≤ Pp best then Pp best ← Pposition;
         if Pcost ≤ Pg best then Pg best ← Pp best;
         end
      end
   end
end
return Pg best;
```

2.17 Multiswarm Optimization

Multiswarm optimization (MSO) is a variant of particle swarm optimization (PSO). Instead of using one swarm, MSO uses a user-defined number of swarms to locate the global optimum (Zhao et al. 2008). The algorithm is especially useful for multimodal optimization problems, where numerous local optima exist. Multiswarm optimization is a new approach to improving the balance between global search and local search.

2.18 Random Search

The random search (RS) algorithm, as the name suggests, is a simple random search algorithm. It takes any position in the search space with equal probability (Brooks 1958). The new solutions are always independent of the previous ones. Random search provides a candidate solution construction and evaluation routine. The main parameters of the algorithm are as follows:

Algorithm RS Pseudocode of Random Search Algorithm (Brownlee 2011)

```
Input: NumIterations                      /*Number of iterations*/
       ProblemSize                        /*Dimension of the search space*/
       Populationsize           /*Size of the population*/
       SearchSpace                        /*Search space of the objective function*/
Output: Best
Best ← ∅;
foreach iteri ∈ NumIterations do
    candidatei ← RandomSolution(ProblemSize, SearchSpace);
    if Cost(candidatei) < Cost(Best) then Best ← candidatei;
    end
end
return Best;
```

2.19 Simulated Annealing

The Simulated Annealing (SA) method was described by Kirkpatrick (1984) and published in 1983. The operation of the algorithm is based on a physical phenomenon. In metallurgy, certain materials gain beneficial properties when heated and then cooled under controlled conditions. The crystal structure of the material is transformed during the process because the particles take more favourable positions. The metaheuristic algorithm emulates this process to search for better solutions to a given problem. Each solution of the algorithm represents an energy status of the system. The Metropolis–Hastings Monte Carlo algorithm controls the acceptance of the new position. As the system is cooled, the acceptance criteria of samples are narrowed to focus on improving movements. The rate of cooling and the definition of the neighbourhood are two important parameters. The cooling rate values applied in the literature differ from study to study. To select the proper value, some preruns are needed. The calculation of the temperature is as follows: temp = temp * temp_change.

The size of the neighbourhood considered in generating candidate solutions may also change over time or be influenced by the temperature, starting initially broad and narrowing with the execution of the algorithm. If the choice of neighbourhood is too small, then the resulting simulated process will not be able to move around the feasible region quickly enough to reach the minimum in a reasonable time. On the other hand, if the neighbourhoods are too large, then the process essentially performs a random search through the feasible region. Intuitively, it seems that a neighbourhood system that strikes a compromise between these extremes would be best.

The definition of the neighbourhood for combinatorial problems is the next permutation. For continuous problems, the neighbourhood is represented as a movement along the variable(s) in the design space.

The algorithm implementation uses a two-opt ($k = 2$) procedure for the neighbourhood function and the classical P(accept) as the acceptance function (Brownlee

2011; Eglese 1990). According to Goldstein and Waterman (1988), the number of k neighbours of any given point in the feasible region is approximately $\binom{N}{k}^{\frac{k-1}{2}} 2^k$.

The common acceptance method is always to accept improving solutions and accept worse solutions with a probability of $P(\text{accept})) \leftarrow \exp((e - e')/T)$, where T is the current temperature, e is the energy (or cost) of the current solution, and e' is the energy of a candidate solution being considered. A simple linear cooling regime is used with a large initial temperature, which is decreased in each iteration.

Theoretically, if the cooling process is long enough, the system always converges to the global optimum. The continuous version of the algorithm was developed by Corana et al. (1987). Attempts have been made to improve its performance (Hasançebi et al. 2010).

The main parameters of the algorithm are as follows:

Algorithm SA Pseudocode of Simulated Annealing Algorithm (Brownlee 2011)

```
Input: ProblemSize        /*Dimension of the search space*/
       iterationsmax      /*Number of iterations*/
       tempmax            /*Maximal temperature*/
Output: Sbest
Scurrent 1 ← CreateInitialSolution(ProblemSize);
Sbest ← Scurrent;
for i = 1 to iterationsmax do
   Si ← CreateNeighborSolution(Scurrent);
   tempcurr ← CalculateTemperature(i, tempmax);
   if Cost(Si) ≤ Cost(Scurrent) then Scurrent ← Si;
      if Cost(Si) ≤ Cost(Sbest) then Sbest ← Si;
      end
   else if Exp( (Cost(Scurrent) − Cost(Si)) / tempcurr ) > Rand() then Scurrent ← Si;
   end
end
return Sbest;
```

Table 2.1 shows the tuning parameters for the different algorithms. The values given are usually good, but in some cases, when the number of unknowns becomes higher, and the constraint is highly nonlinear, some modifications are needed.

2.20 Practical Examples

In this section, we will look in-depth at particle swarm optimization and artificial bee colony algorithms. After introducing the mathematical model, we will implement both of the algorithms in MATLAB with a description for each line of code.

Table 2.1 Parameter set and values for different algorithms

Algorithm	Name of the parameter	Value of the parameter
Bacterial foraging algorithm (BFOA)	Cellsnum	50
	Ned	4
	Nre	5
	Nc	25
	Ns	10
	Stepsize	0.05
	Ped	0.25
Bat algorithm (BATA)	PopulationSize	40
	Loudness	0.5
	PulseRate	0.5
	Frequency	2
Bees algorithm (BA)	Beesnum	50
	Sitesnum	3
	EliteSitesnum	1
	PatchSizeinit	3
	OtherBeesnum	2
	EliteBeesnum	7
Cross-entropy method (CEM)	Samplesnum	50
	UpdateSamplesnum	5
	Learnrate	0.7
Cuckoo search algorithm (CS)	NestSize	40
	DiscoveryRate	0.25
Cultural algorithm (CA)	Populationnum	50
Differential evolution (DE)	Populationsize	50
	Weighting factor	0.8
	Crossover rate	0.9
Firefly algorithm (FF)	PopulationSize	40
	Gamma	1
Harmony search (HS)	Memory size	50
	Consolidationrate	0.95
	PitchAdjustrate	0.7
Memetic algorithm (MA)	Population size	50
	MemePopsize	16
Multiswarm optimization (MSO)	NumberOfSwarms	3
	c3	0.3645
	death	0.005

(continued)

Table 2.1 (continued)

Algorithm	Name of the parameter	Value of the parameter
	immigrate	0.005
Nelder–Mead algorithm (NM)	Amoeba size	20
Random search (RS)	Population size	50
Particle swarm optimization (PSO)	Populationsize	50
	W weighting factor	0.729
	C1 cognitive learning coefficient	1.49445
	C2 social learning coefficient	1.49445
Simulated annealing (SA)	temp_change	0.995
	Tempmax	Searchspace (max)-Searchspace (min)

*Search space (min) (max) means the min and max values of the variables

2.20.1 Particle Swarm Optimization (Eberhart and Kennedy 1995)

We can refer to the members of the swarm and the swarm itself as the particle and population, respectively, and every particle is a candidate solution to the optimization problem that should be solved. The search space limits all the possible solutions to a problem, and the particles have to reach to the best position (the best solution for the optimization problem) in the space. The position and velocity of a specific particle are denoted by $x_k(t) \in S$ and $v_k(t) \in S$, respectively, where k is the index of the particle in the swarm and S is the search space, while (t) is a discrete time step that shows the iteration number of the algorithm. The velocity and position vectors are located in the same space with the same dimensionality, while $(t + 1)$ indicates the next time step. Each particle has its own experience and its own memory about the best position where it was—we shall call it personal best of the kth particle, and it is denoted by $p_k(t)$. The particles communicate and interact with each other and share their personal experience, so they learn and decide what is the best experience among all the experiences of the other party—we shall call this global best, and it is denoted by $G(t)$. The following quantities express the PSO mathematical model:

$$x_{kj}(t+1) = x_{kj}(t) + v_{kj}(t+1), \tag{2.16}$$

$$v_{kj}(t+1) = w * v_{kj}(t) + r_1 C_1(p_{kj}(t) - x_{kj}(t)) + r_2 C_2(G_j(t) - x_{kj}(t)), \tag{2.17}$$

$v_{kj}(t + 1)$: the velocity of particle k in time step $(t + 1)$ and the jth component for this velocity

r_1, r_2: a random number uniformly distributed in the range 0–1

C_1, C_2: the acceleration coefficient
$w * v_{kj}(t)$: inertia term
w: inertia coefficient
$r_1 C_1(p_{kj}(t) - x_{kj}(t))$: the cognitive component
$r_2 C_2(G_j(t) - x_{kj}(t))$: social component.

Equations (2.9) and (2.10) are the two rules that should be followed by all particles in the swarm, and that is the exact meaning of the swarm intelligence. By defining these rules on every iteration of PSO, the velocity and position of each particle are updated according to this simple mechanism.

To implement the PSO algorithm, consider the following minimization problem:

$$y = \sum_{i=1}^{n} x_i^2, \qquad (2.18)$$

where $i = 1 \ldots n$ and $x = [x_1, x_2, x_3, x_4]$.

This is the sphere optimization function, whose global minimum is zero at $x = [0, 0, \ldots, 0]$. First of all, create a new script file and save it in a specific location on the hard disk as PSO.m. Write the commands:

```
clear;
clc;
```

The *clear* command is to delete all variables which may be initiated from a previous session and may cause conflict with the current work, while the *clc* command is just for cleaning the command window. It is better to write a separate function file for the cost function shown in Eq. (2.18); so create another script file and write the code:

```
function out = cost(x)
out= sum(x.^2);
end
```

The function file should start with the keyword *function* and end with the keyword *end*. The variable `out` represents the output of the function, and `cost` is the name of the function. (x) represents the input of the function; it must be located between two parentheses. We are planning to send a vector consisting of four variables. Thus, it is necessary to use a dot before power sign to ensure rising each element in the vector x to the power of 6. After raising every element to the power of 2, the built-in function sum makes the summation for that element, and this is the output of the function that is stored in variable out. Do not forget to end the statement with a semicolon to prevent the variable out from appearing on the command window; this does not affect the work of the algorithm.

The second step is defining the problem parameters. Write the code:

```
%% problem parameters
costfun = @(x) cost(x);
nvar = 4;
V_size = [1,nvar];
lowerLimit = -20;
upperLimit = 20;
```

The double percentage sign `%%` starts a new section and is intended for the parameters of the problem. The function handle `costfun` holds an anonymous function that receives a vector (x) and sends it to the cost function. This function is called whenever there is a need to calculate the cost of the vector of the variables. The number of variables `nvar` represents how many variables govern the optimization problem, while v_size is a row vector representing the size of the variables (we shall deal with it later). The limits of the search space should be defined by the variables `lowerLimit` and `upperLimit`, which are the lower limit and upper limit of the search space.

Start a new section for the parameters of the PSO:

```
%% PSO parameters
nIter = 200;
nPop = 50;
w = 1;
c1 = 2; % personal acceleration
c2 = 2; % social acceleration
damping = 0.95;
```

Particle swarm optimization is an iterative algorithm, so we have to define the variable `nIter`, which is the number of iterations. The next parameter is `nPop`, which represents the number of agents in the party we have explained that every agent or particle is a candidate solution for the problem. The inertia coefficient, as well as personal and social acceleration coefficients w, $c1$, and $c2$, respectively, is explained in Eq. (2.17). We shall talk about the damping variable later. Start a new section, and call it initialization to initialize a starting point for the particles in the search space. Each particle has a position and velocity according to the mathematical model of the PSO, so we have to define a structured form for the particles and assign position and velocity as properties.

```
%% initialization
% template
solution_templet.position = [];
solution_templet.velocity = [];
```

Each particle has its own measurement, and it should be added to the structure as a cost property.

```
solution_templet.cost = [];
```

Particles have a memory about their best experience—where the best location was and how much the cost was for that location—so add the best property to the template, and it is also a structure for the position and cost:

```
solution_templet.best.position = [];
solution_templet.best.cost = [];
```

For all 50 particles, the solution template should be repeated. The *repmat* function is appropriate for this problem, and the function has the syntax (matrix, dimension) so in the following code (a 50×1 column vector), each element is a structure of the above-mentioned properties.

```
% create particles
particle = repmat(solution_templet, 50,1);
```

The best experience among all the experiences of the particles, or say the global best $G(t)$ should be started with the worst value and updated during the progress of the algorithm. For minimization problems, the worst value is Inf, while for maximization problems, the worst value is—Inf:

```
Globalbest.cost = Inf; % start with the worst value
```

Let us start with initializing a random value of the properties of the particles, i.e. we have to make a loop repeated many times as the number of the particles:

```
% initialize population
for i=1:nPop
```

Start a random position for the particles using continuous uniform random number function and assign lower and upper limits as the first and second arguments. The third argument represents the dimension of the required random numbers, assigning variable size `V_size` return 1×4 vector of random numbers between the lower and upper limits. This vector represents the position of the particle; for this problem, the position is represented by four dimensions because there are four governing variables for the cost function:

```
particle(i).position =
unifrnd(lowerLimit,upperLimit,V_size);
```

Initially, the velocities are assumed to be zeros for all the particles.

```
particle(i).velocity = zeros(V_size);
```

Use the function handler to calculate the cost corresponding to the position of each particle.

```
particle(i).cost = costfun(particle(i).position);
```

At this moment, there is no best experience for the particles other than its current experience (the initial one), so we can equalize the personal best position and cost to the initial position and cost.

```
% update personal best
particle(i).best.position = particle(i).position;
particle(i).best.cost = particle(i).cost;
```

The global best can be updated by making a comparison with the personal best. If the *i*th personal best cost is less than the current global best, then replace the global with this particle:

```
    % update global best
    if (particle(i).best.cost < Globalbest.cost)
        Globalbest = particle(i).best;
    end
end
```

For purposes of reporting or analysis, we may define a variable to hold all the best costs over iteration. This does not affect the working of the algorithm, but it is useful when you study the performance of the algorithm:

```
% array to hold best costs value for each iteration
BestCost_Value = zeros(nIter,1);
```

Now, we can update the positions and velocities of the particles, so start a new section for the main loop:

```
%% main loop
```

The main loop consists of two nested loops; the first loop is the iteration loop, which is repeated for the number of iterations:

```
for it =1:nIter
```

At each iteration, the position and velocity of the particle are updated. Thus, there should be a loop repeated as many times as the number of populations:

```
    for i=1:nPop
```

Apply Eq. (2.17) to calculate the velocity of the ith particle. Note that the acceleration coefficients $c1$ and $c2$ are constants in every loop, so they are defined in the section of parameters of the PSO, while r_1 and r_2 [which are described in (2.17)] are random numbers generated at each loop and they are represented by `rand(V_size)`. This argument returns a 1×4 vector of random numbers:

```
        particle(i).velocity =
w*particle(i).velocity+...
            c1*rand(V_size).*(particle(i).best.position
-                             particle(i).position)+...
            c2*rand(V_size).*(Globalbest.position -
                              particle(i).position);
```

Apply Eq. (2.16)

```
particle(i).position = particle(i).position +
                  particle(i).velocity;
```

So far, every particle has updated its velocity and moved to a new position. Consequently, the cost of the new position has to be evaluated; thus, send the new position of the ith particle to the function handler `costfun` to calculate the cost for that new position:

```
particle(i).cost = costfun(particle(i).position);
```

The personal best has to be updated by comparing the best cost of the particle with the newly evaluated cost for the new position. If the cost is less than the personal best cost, then replace the old personal best position with the new position and also replace the old best cost with the cost of the new particle:

```
        % update personal best
            if(particle(i).cost < particle(i).best.cost)
                particle(i).best.position =
particle(i).position;
                particle(i).best.cost = particle(i).cost;
```

If the personal best cost is less than the global best cost, then replace the global best with the new personal best. This is how to update the global best:

```
                % update global best
                if(particle(i).best.cost < Globalbest.cost)
                    Globalbest = particle(i).best;
                end

            end

    end
```

For a specific iteration step, add the global best cost of the vector:

```
BestCost_Value
    BestCost_Value(it) = Globalbest.cost;
```

Display the information about the iteration and corresponding best cost value among all the costs of the particles:

```
    disp(['iteration' num2str(it) ': Best cost = '
num2str(BestCost_Value(it))]);
```

Up to now, the algorithm should work, but if we use a constant inertia coefficient `w`, the convergence will be very little and unacceptable, so we have to reduce the value of `w` by a small factor. This is done by multiplying `w` by the `damping` variable as follows:

```
    w = w*damping;
end
```

2.20.2 *Artificial Bee Colony*

The model of the ABC algorithm is based on the following:

Scout Bees Phase (Initialization)

Repeat

Employed bees' section

Onlooker bees' section

Scout bees' section

Memorize the best solution in the current trail

Until(stop condition reached; maximum numbers of cycles)

Each section of the algorithm has its own low-level structure, and these structures affect global level by interactions between them. Initially, all bees are scouts and search for new solutions randomly. Assume x is a vector of random solutions that are initially returned by scout bees

$$x = (x_1, x_2, x_i \ldots, x_{n-1}, x_n), \tag{2.19}$$

where $n \in R^n$ $i = 1\ldots n$.

2.20.2.1 Employed Bees' Section

These are foragers associated with a particular food source, and they have information about the source that they are exploiting. Employed bees share their information with other bees about the direction, distance, and richness of the food source. Reference (Karaboga 2010) proposes the following formula:

$$v_i = x_i + \phi_i (x_i - x_k), \tag{2.20}$$

where v_i is the new solution vector, ϕ_i is a random number in the period $[-1, 1]$, and k is a random number representing the different random order in the solution vector.

2.20.2.2 Onlooker Bees' Section

These are the bees waiting in the hive and keep watching of the dancing bees that are coming from a particular or randomly discovered food source and extract information from dancers. Bees share their information using waggle dance (Computing 2011), and onlooker bees use probability, to select the best solution. Roulette wheel selection method (Goldberg and Holland 1988) which is a fitness value-based selection technique can be used to find this the probability of a solution (P_i):

$$P_i = \frac{f_i}{\sum_{i=1}^{\text{pop}} f_i} \tag{2.21}$$

$$f_i = \begin{Bmatrix} \frac{1}{1+O_i} & O_i \geq 0 \\ 1 + \text{abs}(O_i) & O_i < 0 \end{Bmatrix} \tag{2.22}$$

where f_i is the fitness value of the objective function O_i.

2.20.2.3 Scout Bees' Section

At the beginning of the algorithm, all the bees are scouts and later convert to employed or onlooker bees during run-time. The employed bees whose position (solution) does not change after a specific number of trails have to abandon their position and convert to scouts. Abandonment criterion, which is called limit control, is very important to jump from the local minimum and continue to search for the global minimum of the optimization problem.

2.20.2.4 Implementation of ABC

Consider the Rosenbrock Valley function (Rosenbrock 1960)

$$f(x, y) = (1 - x)^2 + 100(y - x^2)^2. \tag{2.23}$$

It is a difficult optimization problem with many local minima points, and its global minimum value is zero at $x = 1$ and $y = 1$.

Many of the programming principles are re-repeated. Thus, we have to skip some of the basics explained in the previous sections. Let us start coding, write separated function file which receives one input and return one output for Rosenbrock's Valley function that should be optimized. Note that the input is a vector of two quantities [see Eq. (2.23)], so we assume the first variable is the x-value and the second variable is the y-value.

```
function fxy = Rosenbro(s)
x = s(1);
y = s(2);
fxy =(1-x)^2+100*(y-x^2)^2;
end
```

Save the function file and start the new script with the following code, and save it as ABC.m:

```
clc;
clear;
close all;
```

Define the parameters of the problem:

```
%% Problem Definition
CostFunction=@(x) Rosenbro(x);
```

The number of variables is as same as the number of variables of the objective function.

```
nVar=2;
VarSize=[1 nVar];
lowerLimit =-10;
upperLimit = 10;
```

Define the parameters of the artificial bee colony algorithm where the number of the employed bees is equal to the number of the onlooker bees, which is half of the population size. We have to assume a significant value for the limit control.

```
%% ABC parameters
MaxIt=300;                  % Maximum Number of Iterations
nPop=100;                   % Population Size
employed = nPop/2;          % Number of employed Bees
nOnlooker=nPop/2;           % Number of Onlooker Bees
```

Employed bees should be converted to scouts if their solutions do not improve after a predetermined number of trails called the limit. Here we set a large value for the limit parameter.

```
Limit = nOnlooker*nPop;         % abandonment criteria
```

Initialize a structure bee with position and cost properties, and repeat it to generate the population of the hive:

```
%% Initialization
templete_bee.Position=[];
templete_bee.Cost=[];
Bee = repmat(templete_bee,nPop,1);
```

The optimization problem in this example is a minimization problem, so the worst value that should start for the cost of the best solution is Inf.

```
BestSol.Cost=Inf;
```

Initialize the population, and determine the cost of each member of the population.

```
for i=1:nPop
Bee(i).Position=unifrnd(lowerLimit,upperLimit,VarSize);
Bee(i).Cost=CostFunction(Bee(i).Position);
```

Replace the initial value of the best solution variable with the best solution ever found in the population.

```
    if Bee(i).Cost<=BestSol.Cost
        BestSol=Bee(i);
    end
end
```

Each member of the population is a candidate for a specific solution, so for each of them we have to define a counter for the abandonment criterion, and then define a variable for the best cost over each trial.

```
% Abandonment Counter
C=zeros(nPop,1);
% Array for Best Cost Values
BestCost=zeros(MaxIt,1);
```

Start a new section for the main loop, which consists of three subsections.

```
%% ABC Main Loop
```

The main loop usually stops when the number of cycles reaches the maximum number of iterations, which is the stop criterion of the algorithm.

```
for it=1:MaxIt
```

The first section is updating the position of the employed bees:

```
        for i=1:employed
```

The random number ϕ_i in Eq. (2.20) was taken to be between -1 and 1, while k was taken to be a random integer number between 1 and number of the employed bees.

```
        phi=unifrnd(-1,+1,VarSize);
        k = randi(employed);
```

Use Eq. (2.20) to calculate a new position for the employed bees

```
newbee.Position=Bee(i).Position+phi.*(Bee(i).Position-
Bee(k).Position);
```

The cost of the new position has to be evaluated:

```
newbee.Cost=CostFunction(newbee.Position);
```

Greedy selection should be made between the cost of the new position and the current cost of the position of the employed bee. If the new solution is better than the current solution, then replace the current position with the new position; otherwise the current position of the bee is still as it is, but the abandonment criterion for the position has to be increased by 1.

```
        if newbee.Cost < Bee(i).Cost
            Bee(i)=newbee;

        else
            C(i)=C(i)+1;
        end

    end
```

Define a variable to hold the fitness values of the position of employed bees, start with zero value, and update them using Eq. (2.22).

```
F=zeros(nPop,1);
for a=1: nPop
    if(Bee(a).Cost>=0)
        F(a)= 1/(1 + Bee(a).Cost);
    else
        F(a) = 1+ abs(Bee(a).Cost);
    end
end
```

Probability P can be estimated based on the previous steps and Eq. (2.21). Axiomatically, the values of the vector P range from 0 to 1.

```
P = F/sum(F);
for m=1:nOnlooker
```

Use the Roulette wheel selection function to find the best order in the employed bees according to the probability.

```
        i=RouletteWheelSelection(P);
```

The code of the function Roulette wheel selection should be written in a separate script file and saved as `RouletteWheelSelection.m`

```
function i=RouletteWheelSelection(P)
```

Generate a random number between 0 and 1:

```
    r=rand;
```

Calculate the cumulated summation of the vector P.

```
    C=cumsum(P);

    i=find(r<=C,1);

end
```

Go back to the main loop and repeat the same procedure as in the employed section to calculate a new solution using Eq. (2.20):

```
k = randi(nPop);
phi=unifrnd(-1,+1,VarSize);
newbee.Position=Bee(i).Position+phi.*(Bee(i).Position-
Bee(k).Position);
```

Estimate the objective function value for the new solution and update the solution. If the solution has not changed, increase the abandonment counter.

```
newbee.Cost=CostFunction(newbee.Position);
        if newbee.Cost<=Bee(i).Cost
            Bee(i)=newbee;
        else
            C(i)=C(i)+1;
        end
    end
```

The final section in the main loop is the scout bees. For all bees, check whether the abandonment counter is equal to the abandonment criterion. If so, then convert the bee to a scout and randomly search for a new solution in the search space, and reset the abandonment counter of the newly converted bee to zero:

```
        for i=1:nPop
        if C(i)>=Limit
Bee(i).Position=unifrnd(lowerLimit,upperLimit,VarSize);
            Bee(i).Cost=CostFunction(Bee(i).Position);
            C(i)=0;
        end
    end
```

Compare the best solutions of the bees with the random solution given by scouts and make a greedy selection.

```
    for i=1:nPop
        if Bee(i).Cost<=BestSol.Cost
            BestSol=Bee(i);
        end
    end
```

Add the cost of the best solution of the bees to the array of the best costs to draw on later for analysis purposes and tracking the efficiency of ABC.

```
        BestCost(it)=BestSol.Cost;
```

Display iteration information:

```
disp(['Iteration ' num2str(it) ': Best Cost = '
num2str(BestCost(it))]);
```

End the iteration loop.

```
end
```

Plot best costs against iterations:

```
%% Results
figure;
%plot(BestCost,'LineWidth',2);
semilogy(BestCost,'LineWidth',2);
xlabel('Iteration');
ylabel('Best Cost');
grid on;
```

The following code is for the Roulette wheel function:

```
function a=RouletteWheelSelection(prob)
    k=rand;
    C=cumsum(prob);
    a=find(k<=C,1);
end
```

2.21 Summary

The optimization algorithm field is one of the main areas of computational intelligence which is used to solve a wide range of engineering applications. This chapter introduces the most known algorithms with short explanations about its origin and implementation. In many cases, we deal with Robotic problems as an optimization problem which is desired to find the minimum or maximum of a quantity. In the following chapters, we will learn the formulations of objective functions for different robotic problems.

References

Brooks SH (1958) A discussion of random methods for seeking maxima. Oper Res 6(2):244–251

Brownlee J (2011) Clever algorithms: nature-inspired programming recipes. Available online: https://doc.lagout.org/science/0_Computer%20Science/2_Algorithms/Clever%20Algorithms_%20Nature-Inspired%20Programming%20Recipes%20%5BBrownlee%202012-06-15%5D.pdf. Accessed 18 Dec 2019

Carbas S (2016) Design optimization of steel frames using an enhanced firefly algorithm. Eng Optim 48(12):2007–2025

Colorni A, Dorigo M, Maniezzo V (1991) Distributed optimization by ant colonies. In: Proceedings of the first European conference on artificial life. Paris, France, pp 134–142

Colorni A, Dorigo M, Maniezzo V (1992) An investigation of some properties of an "ant algorithm". In: PPSN, pp 509–520

Computing GTCO (2011) The waggle dance of the honeybee. Youtube. https://www.youtube.com/watch?v=IkO1uJeOlvM. Accessed 4 Aug 2018

Corana A, Marchesi M, Martini C, Ridella S (1987) Minimizing multimodal functions of continuous variables with the "simulated annealing" algorithm Corrigenda for this article is available here. ACM Trans Mathem Softw (TOMS) 13(3):262–280

Dorigo M (1992) Optimization, learning and natural algorithms. PhD Thesis, Politecnico di Milano

Dorigo M, Gambardella LM (1997) Ant colony system: a cooperative learning approach to the travelling salesman problem. IEEE Trans Evol Comput 1(1):53–66

Eberhart R, Kennedy J (1995) A new optimizer using particle swarm theory. In: Micro machine and human science. In: Proceedings of the sixth international symposium on MHS'95. IEEE, pp 39–43

Eglese R (1990) Simulated annealing: a tool for operational research. Eur J Oper Res 46(3):271–281

Geem ZW, Kim JH, Loganathan GV (2001) A new heuristic optimization algorithm: harmony search. Simulation 76(2):60–68

Goldberg DE, Holland JH (1988) Genetic algorithms and machine learning. Mach Learn 3(2):95–99

Goldstein L, Waterman M (1988) Neighbourhood size in the simulated annealing algorithm. Am J Mathem Manag Sci 8(3–4):409–423

Hasançebi O, Çarbaş S, Saka MP (2010) Improving the performance of simulated annealing in structural optimization. Struct Multidiscip Optim 41(2):189–203

Heris SMK (2018) ACO for continuous domains in MATLAB. Yarpiz. http://yarpiz.com/67/ypea104-acor. Accessed 11 Aug 2018

Hillier MS, Hillier FS (2003) Conventional optimization techniques. In: Evolutionary optimization. Springer, pp 3–25

Karaboga D (2010) Artificial bee colony algorithm. Scholarpedia 5:6915

Kennedy J (2011) Particle swarm optimization. In: Encyclopedia of machine learning. Springer, pp 760–766

Kennedy J, Eberhart R (1995) Particle swarm optimization. In: Proceedings of IEEE international conference on neural networks IV, pp 1942–1948

Kirkpatrick S (1984) Optimization by simulated annealing: quantitative studies. J Stat Phys 34(5–6):975–986

Kong M, Tian P (2006) A new ant colony optimization applied for the multidimensional knapsack problem. In: Asia-Pacific conference on simulated evolution and learning. Springer, pp 142–149

Liu Y, Passino K (2002) Biomimicry of social foraging bacteria for distributed optimization: models, principles, and emergent behaviors. J Optim Theory Appl 115(3):603–628

Mehrabian AR, Lucas C (2006) A novel numerical optimization algorithm inspired from weed colonization. Ecol Inform 1(4):355–366

Mortazavi A, Toğan V (2016) Simultaneous size, shape, and topology optimization of truss structures using integrated particle swarm optimizer. Struct Multidiscip Optim 54(4):715–736

Moscato P (1989) On evolution, search, optimization, genetic algorithms and martial arts: towards memetic algorithms. Caltech concurrent computation program, C3P Report 826

Nelder JA, Mead R (1965) A simplex method for function minimization. Comput J 7(4):308–313

Pham DT, Ghanbarzadeh A, Koç E, Otri S, Rahim S, Zaidi M (2006) The bees algorithm—a novel tool for complex optimisation problems. In: Intelligent production machines and systems. Elsevier, pp 454–459

Reynolds RG (1994) An introduction to cultural algorithms. In: Proceedings of the third annual conference on evolutionary programming. World Scientific, pp 131–139

Rosenbrock H (1960) An automatic method for finding the greatest or least value of a function. Comput J 3(3):175–184

Rubinstein RY (1997) Optimization of computer simulation models with rare events. Eur J Oper Res 99(1):89–112

Socha K, Dorigo M (2008) Ant colony optimization for continuous domains. Eur J Oper Res 185(3):1155–1173
Srikanth GU, Geetha R (2018) Task scheduling using ant colony optimization in multicore architectures: a survey. Soft Comput 22(15):5179–5196
Storn R, Price K (1995) Differential evolution-A simple and efficient adaptive scheme for global optimization over continuous spaces. Technical report, International Computer Science Institute 11 (1995)
Tsutsui S (2007) Cunning ant system for quadratic assignment problem with local search and parallelization. In: International conference on pattern recognition and machine intelligence. Springer, pp 269–278
Whitley LD, Starkweather T, Fuquay DA (1989) Scheduling problems and traveling salesmen: The genetic edge recombination operator. In: ICGA, pp 133–140
Yang XS (2009) Firefly algorithms for multimodal optimization. In: International symposium on stochastic algorithms. Springer, pp 169–178
Yang XS (2010) A new metaheuristic bat-inspired algorithm. In: González JR, Pelta DA, Cruz C, Terrazas G, Krasnogor N (eds) Nature inspired cooperative strategies for optimization (NICSO 2010): studies in computational intelligence, vol 284. Springer, Berlin, Heidelberg, pp 65–74
Yang XS (2012) Flower pollination algorithm for global optimization. In: Durand-Lose J, Jonoska N (eds) Unconventional computation and natural computation. UCNC 2012. Lecture Notes in Computer Science, vol 7445. Springer, Berlin, Heidelberg, pp 240–249
Yang XS, Deb S (2009) Cuckoo search via Lévy flights. In: Nature and biologically inspired computing. World congress on NaBIC 2009. IEEE, pp 210–214
Zhao SZ, Liang JJ, Suganthan PN, Tasgetiren MF (2008) Dynamic multi-swarm particle swarm optimizer with local search for large scale global optimization. In: IEEE congress on evolutionary computation. CEC 2008 (IEEE World Congress on Computational Intelligence). IEEE, pp 3845–3852

Chapter 3
Spatial Representations

3.1 Introduction

The first step in robot kinematics is to establish different coordinate frames attached rigidly to rigid objects. These coordinate frames are essential to represent the position and orientation of a point relative to a reference point. Spatial representation is the basis of other subsequent topics. In this chapter, we study the representation of the position and orientation of a point in space. The notions of orientation matrix and homogenous transformation matrix will be explained, and these notions are the basic factors for subsequent study. A 4×4 homogenous transformation matrix is used to represent both the position and orientation of a rigid body, and this is important when we introduce the concept of forward kinematics in chapter four.

3.2 Position Representation

A point in space is a geometrical entity representing a fixed place in the spatial domain. To describe the position of the point, we need to assign a coordinate frame. The vector connecting the origin of the coordinate frame and the point represents the location of the point in the space relative to that frame. Consider Fig. 3.1, where the vector v_1 is dependent on the coordinate frame and describing point P, which is independent on the coordinate frame $o_1x_1y_1$. The same point P can be represented using another coordinate frame as in Fig. 3.1, where the vector v_2 represents the position of point P concerning the frame $o_2x_2y_2$. Accordingly, v_1 and v_2 are the same as positions of point P concerning frame $o_1x_1y_1$ and $o_2x_2y_2$, respectively. The superscript refers to the reference frame.

$$v_1^1 = P^1 = \begin{bmatrix} 7.57 \\ 13.72 \end{bmatrix}, \; v_2^2 = P^2 = \begin{bmatrix} -4.9 \\ 14.85 \end{bmatrix}$$

H. N. Ghafil and K. Jármai (ed.), *Optimization for Robot Modelling with MATLAB*,
https://doi.org/10.1007/978-3-030-40410-9_3

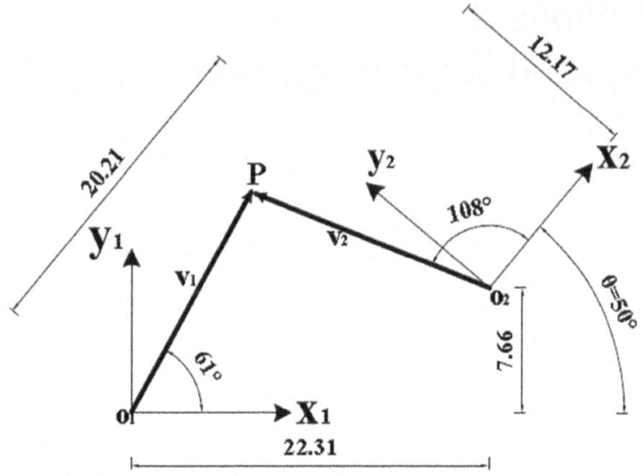

Fig. 3.1 Representation of a point in different coordinate frames

A vector can be represented by magnitude and direction; for instance, v_1 is 15.52 length aligned to x_1-axis by 61°, and v_2 is 15.9 length aligned to x_2-axis by 108°; in this book, we will use an analytical approach to represent vectors by coordinates. In fact, there are infinite numbers of coordinate frames that can represent a single point and it is up to the reader where to locate the origin of a frame, as we will see in chapter four. Also, v_1 and v_2 can be represented concerning to frames $o_1x_1y_1$ and $o_2x_2y_2$, respectively, as follows:

$$v_1^2 = \begin{bmatrix} |v_1|\cos(61-45) \\ |v_1|\sin(61-45) \end{bmatrix} = \begin{bmatrix} 14.918 \\ 4.27 \end{bmatrix},$$

$$v_2^1 = \begin{bmatrix} -|v_2|\cos(180-108-50) \\ |v_2|\sin(180-108-50) \end{bmatrix} = \begin{bmatrix} -14.74 \\ 5.95 \end{bmatrix}$$

v_1^2 and v_2^1 are evaluated under the convention of free vectors, where vectors are free to move in space, and this is appropriate for our work in robotics. The origins o_1 and o_2 are also points in space, and it is possible to represent each one relative to another. Geometrically o_2 locates in the positive direction of x_1 and y_1 while o_1 locates in the negative direction of x_2 and the positive direction of y_2, as shown in Fig. 3.1.

$$o_2^1 = \begin{bmatrix} 22.31 \\ 7.66 \end{bmatrix}, o_1^2 = \begin{bmatrix} -20.21 \\ 12.17 \end{bmatrix}$$

3.3 Rotation Matrix

In vector algebra, the cross product of two vectors gives another vector perpendicular on the two vectors while the dot product of them provides the projection of one on the other. Consider Fig. 3.2, which demonstrates the rotation of frames $o_1x_1y_1$ and $o_2x_2y_2$, which are mentioned in Fig. 3.1 on each other. The axes x_1, y_1, x_2, and y_2 are just vectors that can be represented in another coordinate frame.

In Fig. 3.2a, to represent each point for frame $o_1x_1y_1$ in frame $o_2x_2y_2$, we can say that:

$$x_2 = x_1 \cos(\theta) + y_1 \sin(\theta) \tag{3.1}$$

$$y_2 = -x_1 \sin(\theta)x_1 + y_1 \cos(\theta). \tag{3.2}$$

Equations (3.1) and (3.2) can be written in matrix form:

$$\begin{bmatrix} x_2 \\ y_2 \end{bmatrix} = \begin{bmatrix} \cos(\theta) & \sin(\theta) \\ -\sin(\theta) & \cos(\theta) \end{bmatrix} \begin{bmatrix} x_1 \\ y_1 \end{bmatrix} \tag{3.3}$$

or

$$\begin{bmatrix} x_2 \\ y_2 \end{bmatrix} = R_1^2 \begin{bmatrix} x_1 \\ y_1 \end{bmatrix}, \tag{3.4}$$

where R_1^2 is the rotation matrix of frame $o_1x_1y_1$ concerning frame $o_2x_2y_2$. In the same way, we can write the rotation equation of frame $o_2x_2y_2$ relative to frame $o_1x_1y_1$ for Fig. 3.2b

$$\begin{bmatrix} x_1 \\ y_1 \end{bmatrix} = \begin{bmatrix} \cos(\theta) & -\sin(\theta) \\ \sin(\theta) & \cos(\theta) \end{bmatrix} \begin{bmatrix} x_2 \\ y_2 \end{bmatrix} = R_2^1 \begin{bmatrix} x_2 \\ y_2 \end{bmatrix}, \tag{3.5}$$

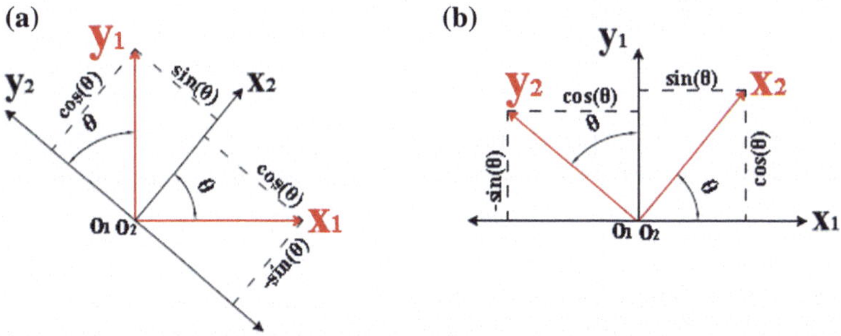

Fig. 3.2 **a** Projection of frames $o_1x_1y_1$ on frame $o_2x_2y_2$ and **b** projection of frames $o_2x_2y_2$ on frame $o_1x_1y_1$

where $(x_1, y_1)^T$ and $(x_2, y_2)^T$ are two vectors representing a point with respect to the frame $o_1x_1y_1$ and $o_2x_2y_2$, respectively. The rotation matrices R_1^2 and R_1^2 in Eqs. (3.4) and (3.5) can be used to map any point from one frame to another. By using the dot product convention, which is just the projection of a vector on another, we can write the rotation matrices in a more general form:

$$R_1^2 = \begin{bmatrix} x_1.x_2 & x_1.y_2 \\ y_1.x_2 & y_1.y_2 \end{bmatrix}, \tag{3.6}$$

$$R_2^1 = \begin{bmatrix} x_2.x_1 & x_2.y_1 \\ y_2.x_1 & y_2.y_1 \end{bmatrix}. \tag{3.7}$$

Now, it is easy to extend the rotation matrix for a three-dimensional case as:

$$R = \begin{bmatrix} x_1.x_2 & x_1.y_2 & x_1.z_2 \\ y_1.x_2 & y_1.y_2 & y_1.z_2 \\ z_1.z_2 & z_1.z_2 & z_1.z_2 \end{bmatrix}. \tag{3.8}$$

Note that the dot product is commutative, i.e., $x_1.x_2 = x_2.x_1$.

3.3.1 *Properties of the Rotation Matrix*

In our example shown in Fig. 3.1, we have mentioned that the angle of rotation between frame $o_1x_1y_1$ and $o_2x_2y_2$ is 50°. It is clear from Fig. 3.2a that the frame $o_1x_1y_1$ rotates 50° in a clockwise direction with respect to frame $o_2x_2y_2$, and from Fig. 3.2b, the frame $o_2x_2y_2$ rotates 50° in a counterclockwise direction. This means that the rotation matrix from Fig. 3.2a is merely the inverse of the rotation matrix from Fig. 3.2b; $\left(R_2^1\right)^{-1}$ is calculated as follows:

$$\left(\begin{bmatrix} \cos(\theta) & -\sin(\theta) \\ \sin(\theta) & \cos(\theta) \end{bmatrix}\right)^{-1} = \begin{bmatrix} \cos(\theta) & \sin(\theta) \\ -\sin(\theta) & \cos(\theta) \end{bmatrix}. \tag{3.9}$$

This means that $R_1^2 = \left(R_2^1\right)^{-1}$.

It is clear from Eq. (3.9) that the inverse of the rotation matrix is just its transpose. Also, the determinant of the rotation matrices and their inverse is 1; thus, we can say that:

1. $R^{-1} = R^T$
2. $R, R^{-1}, R^T \in SO(n)$
3. $\det(R) = 1$

4- Each row and column in R are mutually orthogonal.

Example 3.1 Find the rotation matrix resulting from rotating a coordinate frame 30° about the positive direction of the z-axis.

Usually, the right-hand rule is used to identify the sign of the rotation. Put the thumb in the direction of a vector, and the positive sign rotation will be the direction the rest of the fingers are wrapped in, while the negative sign is in the opposite direction, as shown in Fig. 3.3. Figure 3.4 illustrates the rotation of the frame about its z-axis by $+30°$ and the rotation matrix can be calculated using Eq. (3.8). The projection of the rotated frame $o_1x_1y_1$ and $o_2x_2y_2$ is shown in Fig. 3.4, and the rotation matrix should be

$$R_2^1 = \begin{bmatrix} \cos(\theta) & \sin(\theta) & 0 \\ -\sin(\theta) & \cos(\theta) & 0 \\ 0 & 0 & 1 \end{bmatrix}. \tag{3.10}$$

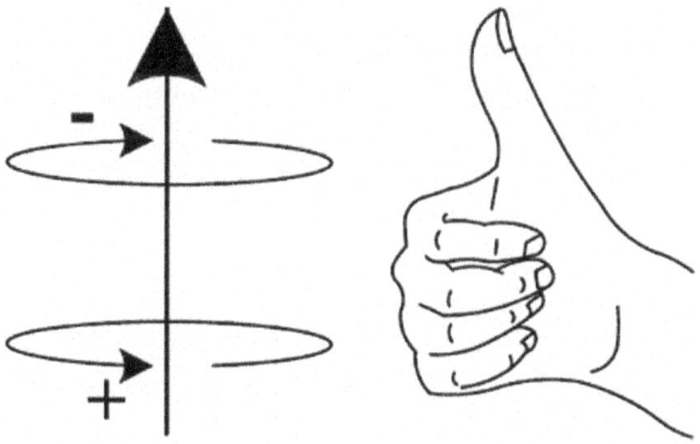

Fig. 3.3 Identifying the sign of rotation using the right-hand rule

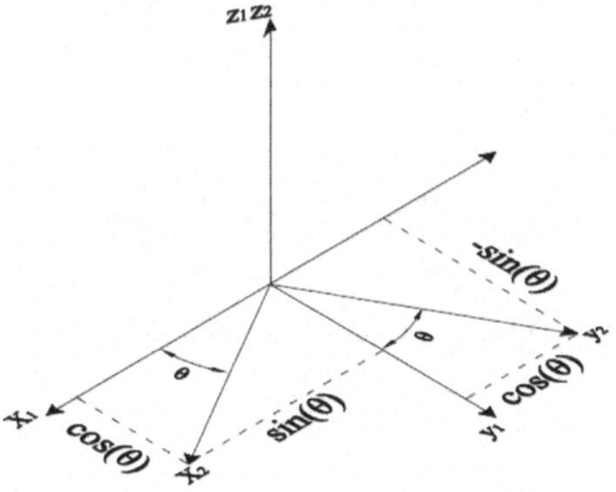

Fig. 3.4 Rotating a coordinate frame about its z-axis

3.4 Composition of Rotations

In robotics, a point is usually subject to many different rotations, and it is necessary to represent the point to a reference frame—we can call it the base frame. Suppose a point P is represented by the coordinate frame $o_3x_3y_3$, as shown in Fig. 3.5a. To represent it relative to the frame $o_2x_2y_2$, we can follow the same procedure for Eqs. (3.4) and (3.5).

$$P^2 = R_3^2 P^3. \tag{3.11}$$

P^3 and P^2 are the representation of point P with respect to frame $o_3x_3y_3$ and $o_2x_2y_2$ while the rotation matrix between the two frames is

$$R_3^2 = \begin{bmatrix} \cos(\phi) & -\sin(\phi) & 0 \\ \sin(\phi) & \cos(\phi) & 0 \\ 0 & 0 & 1 \end{bmatrix}. \tag{3.12}$$

Also, assume that frame $o_2x_2y_2$ makes an angle with frame $o_1x_1y_1$ as in Fig. 3.5b. We can map point P from frame $o_2x_2y_2$ to frame $o_1x_1y_1$

$$P^1 = R_2^1 P^2, \tag{3.13}$$

where

$$R_2^1 = \begin{bmatrix} \cos(\theta) & -\sin(\theta) & 0 \\ \sin(\theta) & \cos(\theta) & 0 \\ 0 & 0 & 1 \end{bmatrix}. \tag{3.14}$$

Substituting Eq. (3.11) into Eq. (3.13) yields

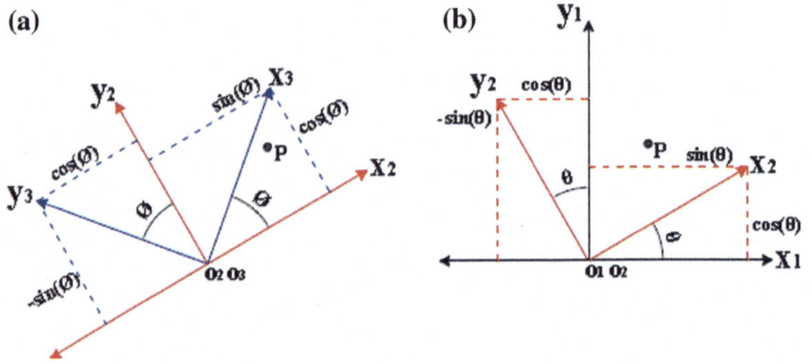

Fig. 3.5 Composition of rotations

$$P^1 = R_2^1 R_3^2 P^3, \tag{3.15}$$

Equation (3.15) is the overall mapping for point P from coordinate frame $o_3x_3y_3$ to the coordinate frame $o_1x_1y_1$ with a composition of rotation:

$$R_3^1 = R_2^1 R_3^2. \tag{3.16}$$

In general,

$$R = \prod_{i=1}^{n} R_{i+1}^{i}, \tag{3.17}$$

where n is the number of rotations.

Multiple rotations about different axes can be considered separately to estimate the final orientation. Assume we have a rotation about x-axis with the angle of θ followed by another rotation about y-axis with the angle of ϕ and then the final rotation matrix can be represented as

$$R = R_{x,\theta}\, R_{y,\phi}. \tag{3.18}$$

3.5 Euler Angles

A rotation matrix can be specified in terms of three independent quantities that are called Euler angles. Figure 3.6 shows the orientation of the frame $o_1x_1y_1z_1$ relative to frame $o_2x_2y_2z_2$, which resulted from successive rotations of three angles (ϕ, θ, ψ) known as Euler angles. They can be used to generate a 3×3 rotation matrix.

Consider Fig. 3.6 where the first rotation of ϕ angle about z_1-axis specifies the orientation of frame (a) with respect to the frame (1)

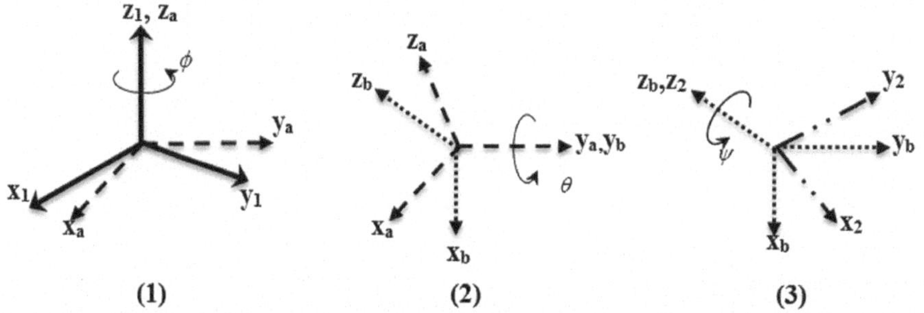

Fig. 3.6 Euler angles

$$R_a^1 = \begin{bmatrix} \cos(\phi) & -\sin(\phi) & 0 \\ \sin(\phi) & \cos(\phi) & 0 \\ 0 & 0 & 1 \end{bmatrix}. \tag{3.19}$$

The second rotation is an angle θ about the current y-axis of the frame (b). Keep in mind that frame (a) resulted from the rotation of angle ϕ, and the rotation matrix of the frame (b) relative to frame (a) will be

$$R_b^a = \begin{bmatrix} \cos(\theta) & 0 & \sin(\theta) \\ 0 & 1 & 0 \\ -\sin(\theta) & 0 & \cos(\theta) \end{bmatrix}. \tag{3.20}$$

The third rotation is an angle ψ about the z-axis of the frame (b) to yield the rotation matrix of frame (2) with respect to frame (b)

$$R_2^b = \begin{bmatrix} \cos(\psi) & -\sin(\psi) & 0 \\ \sin(\psi) & \cos(\psi) & 0 \\ 0 & 0 & 1 \end{bmatrix}. \tag{3.21}$$

From Eq. (3.17), we can find the rotation matrix between frame $o_2x_2y_2z_2$ relative to $o_1x_1y_1z_1$

$$R_2^1 = R_a^1 R_b^a R_2^b, \tag{3.22}$$

$$R_2^1 = \begin{bmatrix} \cos(\phi) & -\sin(\phi) & 0 \\ \sin(\phi) & \cos(\phi) & 0 \\ 0 & 0 & 1 \end{bmatrix} \begin{bmatrix} \cos(\theta) & 0 & \sin(\theta) \\ 0 & 1 & 0 \\ -\sin(\theta) & 0 & \cos(\theta) \end{bmatrix} \begin{bmatrix} \cos(\psi) & -\sin(\psi) & 0 \\ \sin(\psi) & \cos(\psi) & 0 \\ 0 & 0 & 1 \end{bmatrix}. \tag{3.23}$$

In general, R_2^1 is called R_{ZYZ} and denoted as ZYZ Euler angle transformation

$$R_{ZYZ} = \begin{bmatrix} C_\phi C_\theta C_\psi - S_\phi S\psi & -C_\phi C_\theta S_\psi - S_\phi C\psi & C_\phi S_\theta \\ S_\phi C_\theta C_\psi + C_\phi S_\psi & -S_\phi C_\theta S_\psi + C_\phi C\psi & S_\phi S_\theta \\ -S_\theta C_\psi & S_\theta S_\psi & C_\theta \end{bmatrix}. \tag{3.24}$$

where C and S denote cos( ) and sin( ) functions, respectively.

3.6 Roll, Pitch, and Yaw Angles

Other successive rotations about the principal axes of frame $o_1x_1y_1z_1$ called roll, pitch, and yaw angles—denoted as (ϕ, θ, ψ), respectively—can be used to specify the rotation matrix. We will describe the three successive rotations as follows.

The first rotation is roll rotation about the z_1-axis through ϕ angle, then pitch rotation about the y_1-axis by an angle θ, and yaw about the x_1-axis by an angle ψ, as revealed in Fig. 3.7. Since the rotations are made relative to frame $o_1x_1y_1z_1$, as illustrated in Fig. 3.8, the transformation matrix will be

$$R_{XYZ} = \begin{bmatrix} C_\phi & -S_\phi & 0 \\ S_\phi & C_\phi & 0 \\ 0 & 0 & 1 \end{bmatrix} \begin{bmatrix} C_\theta & 0 & S_\theta \\ 0 & 1 & 0 \\ -S_\theta & 0 & C_\theta \end{bmatrix} \begin{bmatrix} 1 & 0 & 0 \\ 0 & C_\psi & -S_\psi \\ 0 & S_\psi & C_\psi \end{bmatrix}$$

$$= \begin{bmatrix} C_\phi C\theta & -S_\phi C_\psi + C_\phi S_\theta S_\psi & S_\phi S_\psi + C_\phi S_\theta C_\psi \\ S_\phi C\theta & C_\phi C_\psi + S_\phi S_\theta S_\psi & -C_\phi S_\psi + S_\phi S_\theta C_\psi \\ S_\theta & C_\theta S_\psi & C_\theta C_\psi \end{bmatrix}. \tag{3.25}$$

Example 3.2 In general, we can write a program using MATLAB to calculate the Euler transformation matrix from the basic three angles. We have considered that the rotation will be in degrees, so we have used the function cosd( ) and sind( ). If the rotations are in radian, then it is proper to use the functions cos( ) and sin( ). Also, the following code is appropriate for both MATLAB and GNU Octave platforms.

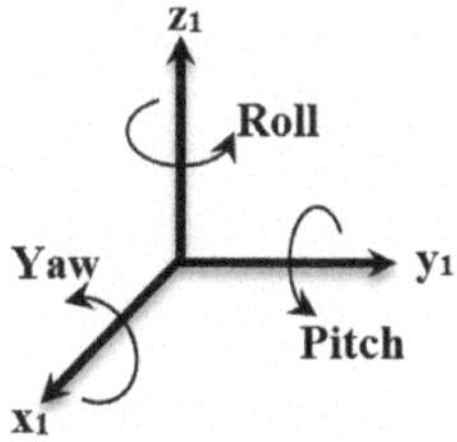

Fig. 3.7 Roll, pitch, and yaw angles

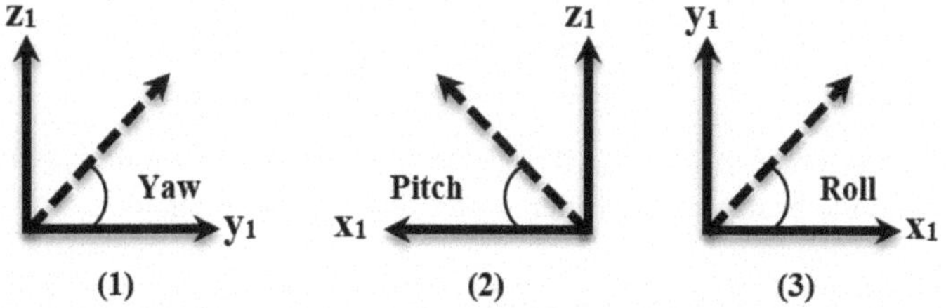

Fig. 3.8 Directions of the roll, pitch, and yaw angles

```
clear;
clc;
phi = input('Enter the value of Phi in degree: ');
theta = input('Enter the value of theta in degree: ');
psi = input('Enter the value of Psi in degree: ');
R_phi = [cosd(phi),-sin(phi),0;...
        sin(phi), cos(phi), 0,;...
         0 , 0, 1];
R_theta = [cosd(theta), 0, sin(theta);...
           0,1, 0;...
         -sind(theta), 0, cosd(theta)];
R_psi = [cosd(psi),-sin(psi),0;...
        sin(psi), cos(psi), 0,;...
         0 , 0, 1];
Euler_AT = R_phi * R_theta * R_psi
```

The same code can be modified to calculate the roll, pitch, and yaw angles using Eq. (3.25), and this will be left to the reader as an exercise.

The more difficult situation is to calculate the Euler angles or roll, pitch, and yaw from the given rotation matrix. We will not explain the complex theoretical equations mentioned in other books because this book is dedicated to using optimization techniques. Thus, the solution is pretty simple from the viewpoint of optimization algorithms. In Sect. 2.20.1, we implemented particle swarm optimization (PSO), and sphere function was introduced as a cost function that should be optimized. Here, we can use the same algorithm with a different cost function and slight modification.

For a given rotation matrix $R_{i,j}$, for $i, j = 1, 2, 3$ its elements will be

$$R = \begin{bmatrix} r_{11} & r_{12} & r_{13} \\ r_{21} & r_{22} & r_{23} \\ r_{31} & r_{32} & r_{33} \end{bmatrix}. \tag{3.26}$$

During the iterative process of the optimization algorithm, there should be optimum Euler angles that give a corresponding rotation matrix $R_{k,\ell}$ such that the difference between each element in $R_{i,j}$ and its corresponding elements in $R_{k,\ell}$ should be zero. Thus, the objective function to find the best Euler angles is

$$f = \sum abs\left(R_{i,j} - R_{k,\ell}\right). \tag{3.27}$$

The following code in MATLAB and Octave is written to implement the cost function to find the best possible Euler angles that can be obtained.

```
function [z] = cost(t)
E = [0.6169,    -0.6604,    -0.6986;...
          0.3358,      0.7003,    -0.8407;...
        -0.3214,     -0.1312,      0.8660];
phi = t(1);
theta= t(2);
psi = t(3);
R_phi = [cosd(phi),-sin(phi),0;...
        sin(phi), cos(phi), 0,;...
         0 , 0, 1];
R_theta = [cosd(theta), 0, sin(theta);...
          0,1, 0;...
        -sind(theta), 0, cosd(theta)];
R_psi = [cosd(psi),-sin(psi),0;...
        sin(psi), cos(psi), 0,;...
         0 , 0, 1];
R = R_phi * R_theta * R_psi;
z= abs(E(1,1)-R(1,1))+ abs(E(1,2)-R(1,2))+ abs(E(1,3)-
R(1,3))+...
  abs(E(2,1)-R(2,1))+  abs(E(2,2)-R(2,2))+  abs(E(2,3)-
R(2,3))+...
  abs(E(3,1)-R(3,1))+  abs(E(3,2)-R(3,2))+  abs(E(3,3)-
R(3,3));
end
```

Matrix (E) is the given rotation matrix while the function has one input, which is a vector of angles representing Euler angles, and one output, which is the summation of the absolute values of the difference between any two elements on the $R_{i,j}$ and $R_{k,\ell}$ that have the same index. We will do the following modification on the PSO parameters.

```
%% problem parameters
costfun = @(x) cost(x);
nvar = 3;
V_size = [1,nvar];
lowerLimit = -120;
upperLimit = 120;
```

The variable `nvar` should be equal to three as it is the same as the number of Euler angles (ϕ, θ, ψ). For this problem, we expect Euler angles in the range of [−120 to 120] and set these values to the `lowerLimit` and `upperLimit` variables. If the reader has no idea about the range of the values of the angles, it is better to set them to the period [−360 to 360]; of course, this will increase the search space. As a result, more iterations and population are needed. For this problem, we proposed 5,000 iterations with a population number of 400.

3.7 Homogenous Transformation Matrix

Rigid motion is defined as a combination of pure rotation with pure translation of a point in space. Refer to Eq. (3.15), where the multiplication of the rotation matrix and position vector was used to transform the vector itself to another coordinate frame. We can concatenate the rotation matrix and position vector in one single 4×4 matrix called homogenous transformation matrix HTM (T) taking the form

$$T = \begin{bmatrix} R & P \\ 0 & 1 \end{bmatrix}, \tag{3.28}$$

Recalling Eq. (3.24), and for a given position vector $P = \left(P_x, P_y, P_z\right)^{\mathrm{T}}$, we can write T as

$$T = \begin{bmatrix} r_{11} & r_{12} & r_{13} & P_x \\ r_{21} & r_{22} & r_{23} & P_y \\ r_{31} & r_{32} & r_{33} & P_z \\ 0 & 0 & 0 & 1 \end{bmatrix}. \tag{3.29}$$

Equation (3.29) is the general form to represent a pure rotation followed by a pure translation of a point. We will explain a convention to drive the homogenous transformation matrix in the next chapter. In order to represent the position and orientation of a rigid body, there should be a coordinate frame attached rigidly to that body. Consider the example shown in Fig. 3.9, where the length of the long link is 20.22 while the length of the shorter one is 14.52. It is clear that frame $o_2x_2y_2z_2$ makes a displacement of 20.22 in the positive direction relative to the x_1-axis and makes no relative rotation to frame $o_1x_1y_1z_1$. Consequently, the rotation matrix is just an identity matrix while the position vector is $(20.22, 0,0)^{\mathrm{T}}$; therefore, the homogenous transformation matrix of the long link will be

$$H_1 = \begin{bmatrix} 1 & 0 & 0 & 20.22 \\ 0 & 1 & 0 & 0 \\ 0 & 0 & 1 & 0 \\ 0 & 0 & 0 & 1 \end{bmatrix}. \tag{3.30}$$

In the same way, we can conclude the HTM for the short link as

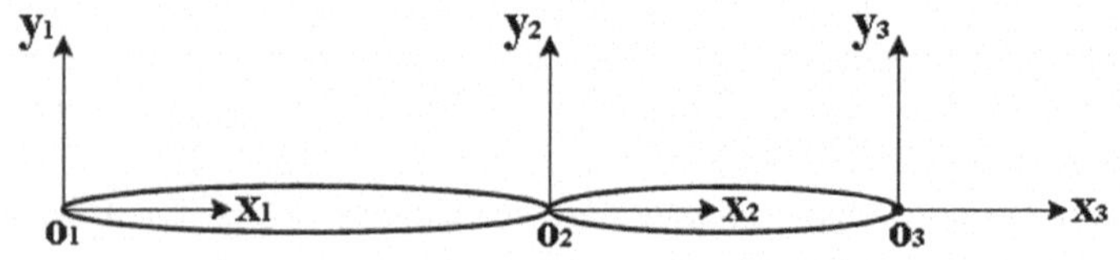

Fig. 3.9 Two rigid bodies in a plane

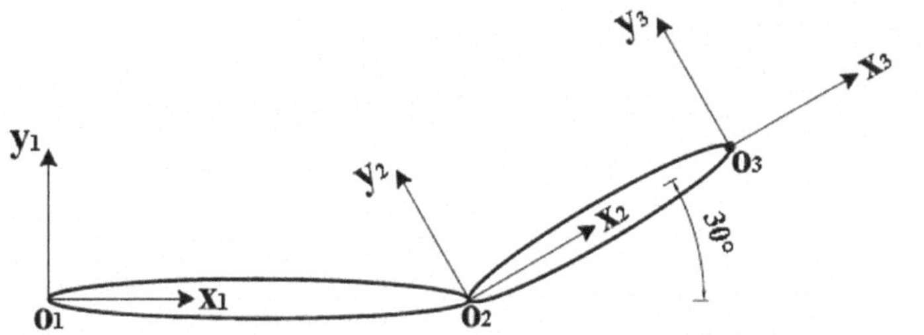

Fig. 3.10 Rotating frame (2) −30° relative to frame (1)

$$H_2 = \begin{bmatrix} 1 & 0 & 0 & 14.52 \\ 0 & 1 & 0 & 0 \\ 0 & 0 & 1 & 0 \\ 0 & 0 & 0 & 1 \end{bmatrix}. \tag{3.31}$$

The overall HTM of both of the two links is the serial multiplication of the two matrices:

$$H = H_1 \cdot H_2 = \begin{bmatrix} 1 & 0 & 0 & 34.74 \\ 0 & 1 & 0 & 0 \\ 0 & 0 & 1 & 0 \\ 0 & 0 & 0 & 1 \end{bmatrix}. \tag{3.32}$$

Indeed, if you measure the distance between points o_3 and o_1, it will be $20.22 + 14.52 = 34.74$ with zero displacement in the direction of the y_1-axis and z_1-axis.

Now, consider rotating frame $o_2x_2y_2z_2$ with ($-30°$) relative to frame $o_1x_1y_1z_1$, as shown in Fig. 3.10. The rotation in a counterclockwise direction is assumed to be negative while the rotation in a clockwise direction is positive; however, the reader can assume the reverse. The position vector of point o_2 did not change while the rotation matrix changes to be like Eq. (3.10). Thus, HTM of the first link will be

$$H_1 = \begin{bmatrix} \cos(-30) & \sin(-30) & 0 & 20.22 \\ -\sin(-30) & \cos(-30) & 0 & 0 \\ 0 & 0 & 1 & 0 \\ 0 & 0 & 0 & 1 \end{bmatrix}. \tag{3.33}$$

The origin point o_3 does not change its relative position or its orientation with respect to frame $o_2x_2y_2z_2$. Thus, the rotation matrix between frame (3) and frame (2) is still the identity matrix with the same position vector as in Eq. (3.31).

The overall HTM of both is

$$H = H_1 \cdot H_2 = \begin{bmatrix} 1 & 0 & 0 & 32.7947 \\ 0 & 1 & 0 & 7.26 \\ 0 & 0 & 1 & 0 \\ 0 & 0 & 0 & 1 \end{bmatrix}. \quad (3.34)$$

Figure 3.10 was drawn, and the distance between o_3 and o_1 was measured. It is 32.79 in x_1-direction and 7.26 in y_1-direction, the same as the results in Eq. (3.34). As shown in Fig. 3.10, when the coordinate frame (2) has rotated by 30°, the whole link with all of its points rotates with the same orientation that is the meaning of coordinate frame attached rigidly to a body.

3.8 Summary

Translational and rotational movement of a point in space can be represented by a single 4×4 homogenous transformation matrix which holds both the position and rotation of a point. This matrix usually used to represent the spatial pose of a rigid body.

Exercises

3.1 Write a code that calculates the rotation matrix given roll, pitch, and yaw angles. You may use the code described in Example 3.2 as a starting point.

3.2 Given the rotation matrix of roll, pitch, and yaw angles

$$R = \begin{bmatrix} 0.2962 & -1.0680 & 0.5335 \\ 0.3379 & 0.0517 & 0.7783 \\ -0.9397 & -0.0453 & 0.3390 \end{bmatrix},$$

write a valid objective function in MATLAB, and use the PSO code in Sect. 2.20.1 to find the optimum values of roll, pitch, and yaw.

Ans ($\phi = -30, \theta = 70, \psi = 25$)

3.3 Try to use artificial bee colony ABC, which is described in Sect. 2.20.2, to optimize the objective function that you have written for Problem 2.

3.4 Could you implement the pseudocode of any of the optimization algorithms described in chapter one?

3.5 Compare the efficiency of PSO and ABC algorithms while they optimize the objective function that you have written for Problem 2. Consider taking the best result among 51 run times.

Chapter 4
Manipulator Kinematics

4.1 Introduction

The kinematics of a robot is the description of the motion of a manipulator without taking consideration of the forces or torques that cause this motion. Within kinematics, time-dependent or independent problems like position, orientation, velocity, and acceleration are studied. In this chapter, we start with an introduction to the manipulator, and then, consider the issues of forward and inverse kinematics. Forward kinematics is finding the end-effector position and orientation given joint variables while inverse kinematics is finding the joint variables given the position and orientation of the end-effector. Forward kinematics is a straightforward operation; in contrast, the inverse kinematic problem is more complicated. By using optimization methods, we will simplify the inverse problem to make it direct for all types of robot manipulators, even those with a high degree of freedom.

4.2 Manipulator

The robot manipulator is a series of mechanical components called links, joined to each other by joints. Usually, manipulators are designed for multi-purpose operations and can be governed directly by a human operator or through a logic device. Figure 4.1 illustrates a virtual model for the famous PUMA manipulator.

Sometimes, the manipulator is called the robot arm or just arm because it consists of many combined pieces in one mechanism, just like the human arm. A robotic prosthetic arm for patients is a clear example of why the manipulator simply arms. There are two main categories of manipulators: the serial manipulator or sometimes called an open-chain robot, where the terminal link ends with an end-effector (e.g. the robots in Figs. 4.1 and 4.3) and the parallel manipulator, where two or more arms share the same ending or end-effector, as shown in Fig. 4.2. This book deals with serial robot manipulators only.

H. N. Ghafil and K. Jármai (ed.), *Optimization for Robot Modelling with MATLAB*,
https://doi.org/10.1007/978-3-030-40410-9_4

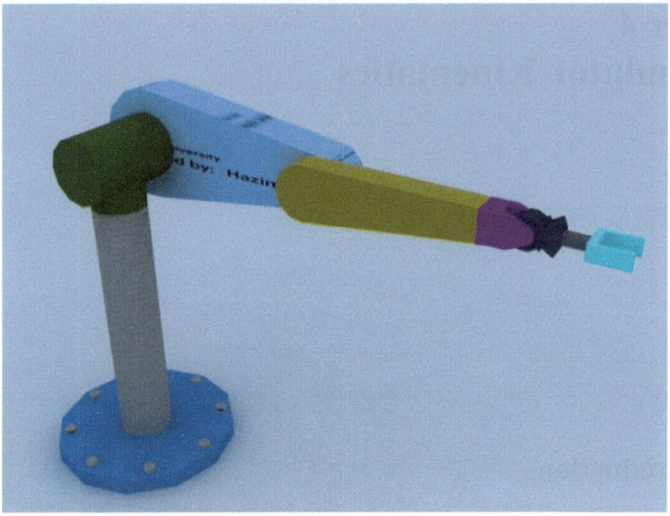

Fig. 4.1 PUMA robot manipulator

Fig. 4.2 Parallel robot manipulator

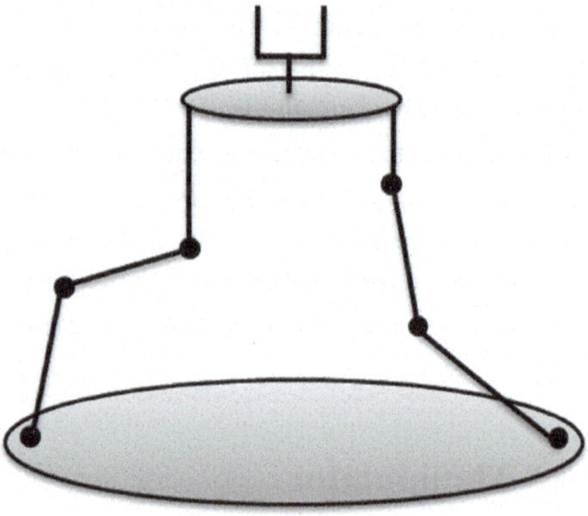

4.2.1 Link

A piece of material, for instance, a rod, beam, etc., is connected with another link by means of joints, as shown in Fig. 4.3. Usually, links are hollow to allow the electrical components to pass inside towards the next link in the chain. Depending on the type of analysis, links can be considered rigid or flexible bodies, as we will explain in subsequent chapters.

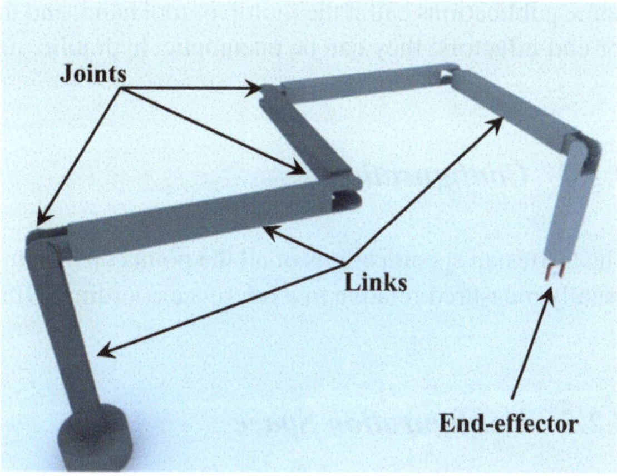

Fig. 4.3 Model of a 7-degrees-of-freedom robot manipulator

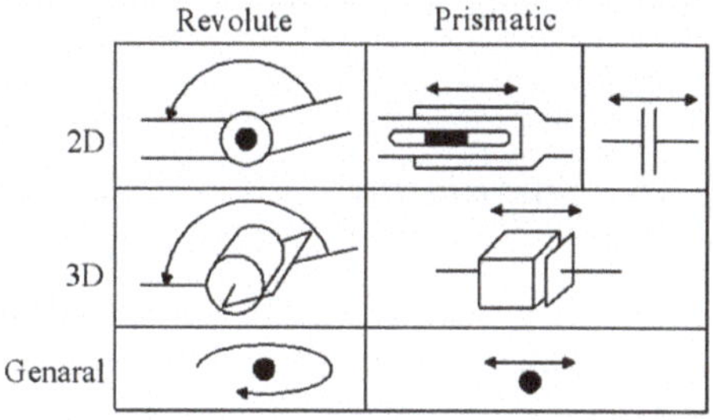

Fig. 4.4 Joint types and representation

4.2.2 Joints

A joint is a physical meeting between two links that allows relative motion between the two. Typically, there are two main types of joints: the revolute joint and the prismatic joint. A revolute joint between two links allows relative rotation between them while a prismatic joint allows relative translation between the two links. Figure 4.4 illustrates revolute and prismatic joints and their representation in 2D and 3D as well as the general representation. Through this book, we will use joint descriptions that are also used in many other books and scientific papers, although there are types like ball-and-socket joints that will not be used in this book.

4.2.3 End-Effector

Figure 4.3 gives a clear image of the placement of the links, joints, and end-effector that form the physical manipulator. The end-effector is the 'hand' of the robot, designed to grasp things or hold machine tools in case of industrial manipulators.

Some publications call it the tooltip or tool hand, and there are different mechanisms for end-effectors; they can be pneumatic, hydraulic, mechanical, or even magnetic.

4.2.4 Configuration

The Cartesian specifications of all the points on a manipulator for a specific pose are usually measured relative to a reference coordinate frame.

4.2.5 Configuration Space

The set of all possible configurations for reachable points within the working area is called the configuration space.

4.2.6 Singularity

Singularity is used to describe the case when the end-effector of a robot manipulator cannot reach a point in the Cartesian space.

4.2.7 Cartesian Space

Cartesian space is a continuous 2D or 3D domain that is restricted by the reachable points for a robot.

4.2.8 Joint Space

Joint space describes a continuous 1D domain constrained by the actuator limits. The numerical values on this space represent the rotation angle for a revolute joint or the translation distance for a prismatic joint.

4.3 Kinematic Chain: Forward Kinematics

The robot manipulator has a degree of freedom equal to the number of joints that connect its links. Each joint has one degree of freedom, rotation for a revolute joint and translational displacement for a prismatic joint. Recalling what we have mentioned in

Sect. 3.7, each link in the arm can be considered a rigid body subject to rigid motion with pure rotation and pure displacement. Consequently, each link can be represented by a single homogenous transformation matrix (HTM) that holds the orientation and position vector of a specified coordinate frame attached rigidly to that link. Each link connects to two joints at the ends, and therefore, for a manipulator who has i links, there should be $i + 1$ joints. The joint variable q_i of the ith joint can be

$$q_i = \begin{cases} \theta_i \text{ For revolute joint} \\ d_i \text{ For prismatic joint} \end{cases}, \tag{4.1}$$

where θ, d are the angle of rotation and linear displacement, respectively. In the same way, as in Sect. 3.7, each link has to be attached rigidly to a coordinate frame so as to achieve the kinematic analysis of the whole robot. Consequently, there is a single HTM that represents the spatial description of the link, and this HTM is a function of the joint variable.

$$H_i = H_i(q_i) \tag{4.2}$$

In robotics, each point on the manipulator links has to be related to a specific coordinate frame called the base frame or reference frame. This frame is the origin of the Cartesian space that the robot works in, and all movements, obstacles, objects, and trajectories should be measured relative to that base frame. Again recall the example in Sect. 3.7: we have related frame (3) to frame (1), so implicitly we have considered frame (1) as a base frame, and the representation of point o_3 with respect to the point o_1 (frame $o_1x_1y_1z_1$) is just the serial multiplication of the homogenous transformation matrices. Consider the serial manipulator in Fig. 4.5. Let us assume that there is a coordinate frame attached rigidly to each link L_i at joint j_i. For this chain, the position and orientation of the end-effector (which lies on the terminal link) relative to the base frame (assume it is located at joint j_1) are just the serial multiplication of the homogenous transformation matrices of all links in the chain.

$$H_n^0 = H_1^0\, H_2^1\, H_n^{n-1}. \tag{4.3}$$

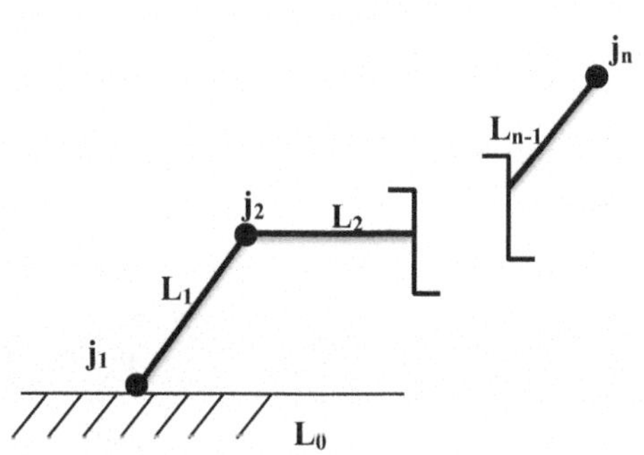

Fig. 4.5 Serial robot manipulator

The overall transformation matrix H_n^0 holds the position and orientation of the point located at a joint (j_n) relative to the base point, which is located at joint j_1. By convention, the coordinate frame that belongs to the position of the tooltip (end-effector) should be found at the central location of the tool; this position is called the tool centre point (TCP). Each joint connects to two links, so a robot manipulator having n joints should have (n – 1) links. Joints are counted starting from (1) while links are numbered starting from (0). Consequently, joint 1 connects link 0 to link 1, and one can say that in general joint (i) connects link (i – 1) to link (i). Consider Fig. 4.5; we will assume that the first link in the manipulator is the ground. Thus, joint (1) connects link (0), which is the ground to link (1), which is the first physical link in the manipulator.

4.4 Denavit–Hartenberg (DH) Convention

The Denavit–Hartenberg (DH) convention is a way to simplify the representation of a point relative to another point. For a robot manipulator with a high degree of freedom, it is not easy to represent the end-effector to the base frame using the traditional dot product of a frame by another frame. The DH convention states that the outcome of four basic transformations can represent every homogenous transformation matrix. Consider Fig. 4.6, the Denavit–Hartenberg convention uses two assumptions to assign coordinate frames to a manipulator:

DH1: Axis x_1 is perpendicular to axis z_0.
DH2: Axis x_1 intersects axis z_0.

The four transformations can be described as follows:

1. Link length (a): the distance between z (i – 1) and z (i).
2. Link twist (α_{i-1}): the angle between z-axes of the link (i – 1) and link i. The direction of the twist angle is performed by the right-hand rule. If you direct your thumb in the direction of $x(i$ – 1) and the rest of your fingers in the direction

Fig. 4.6 Coordinate frames satisfying assumptions DH1 and DH2

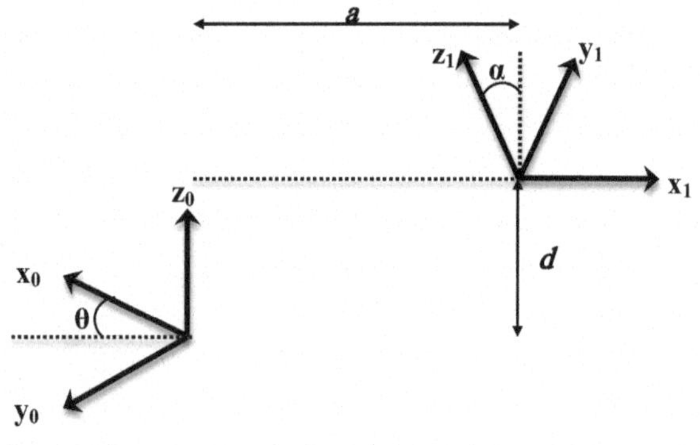

of $z(i - 1)$, then if the $z(i)$ is in the direction of the warp, it will be positive; otherwise, it is negative.

3. Link offset (d_i): the distance between $x(i - 1)$ and $x(i)$.
4. Joint angles (θ_i): the angle between $x(i - 1)$ and $x(i)$. The direction of the joint angle is also allocated by the right-hand rule. Put your thumb in the direction of $z(i - 1)$ and the rest of your fingers in the direction of $x(i - 1)$; then if the joint angle is in the direction of the warp, it will be positive; otherwise, it is negative. The four parameters mentioned above are depicted in Fig. 4.6. With the mechanism of the Denavit–Hartenberg notation available, the process of attaching frames to links to determine the manipulator's forward kinematics is relatively straightforward. In this convention, each homogeneous transformation H is represented as a product of four quantities:

$$H_i = \text{Rot}_{Z,\theta}.\text{Trans}_{Z,d}.\text{Trans}_{x,a}.\text{Rot}_{x,\alpha}$$

$$= \begin{bmatrix} C\theta_i & -S\theta_i & 0 & 0 \\ S\theta & C\theta & 0 & 0 \\ 0 & 0 & 1 & 0 \\ 0 & 0 & 0 & 1 \end{bmatrix} \begin{bmatrix} 1 & 0 & 0 & 0 \\ 0 & 1 & 0 & 0 \\ 0 & 0 & 1 & d_i \\ 0 & 0 & 0 & 1 \end{bmatrix}$$

$$\times \begin{bmatrix} 1 & 0 & 0 & a_i \\ 0 & 1 & 0 & 0 \\ 0 & 0 & 1 & 0 \\ 0 & 0 & 0 & 1 \end{bmatrix} \begin{bmatrix} 1 & 0 & 0 & 0 \\ 0 & C\alpha_i & -S\alpha_i & 0 \\ 0 & S\alpha_i & C\alpha_i & 0 \\ 0 & 0 & 0 & 1 \end{bmatrix}$$

$$= \begin{bmatrix} C(\theta_i) & -S(\theta_i).C\alpha_i & S(\theta_i).S\alpha_i & a_i.C(\theta_i) \\ S(\theta_i) & C(\theta_i).S\alpha_i & -C(\theta_i).S\alpha_i & a_i.S(\theta_i) \\ 0 & S(\theta_i) & C(\theta_i) & d_i \\ 0 & 0 & 0 & 1 \end{bmatrix}. \tag{4.4}$$

It is worth mentioning that the right-hand rule is used to assign the coordinate frames where the thumb is always in the direction of actuation and represents the z-axis of the coordinate frame. The first frame in the manipulator is counted from zero, and the designer is free to choose where to assign a coordinate frame rigidly to a specific link. For a prismatic joint, the rotation angle variable should be equal to zero, and the HTM will be a function of the change in link offset (d_i), which we will denote as (P_i)

$$H_i = \begin{bmatrix} 1 & 0 & 0 & a_i \\ 0 & S\alpha_i & -S\alpha_i & 0 \\ 0 & 0 & 1 & P_i \\ 0 & 0 & 0 & 1 \end{bmatrix}. \tag{4.4a}$$

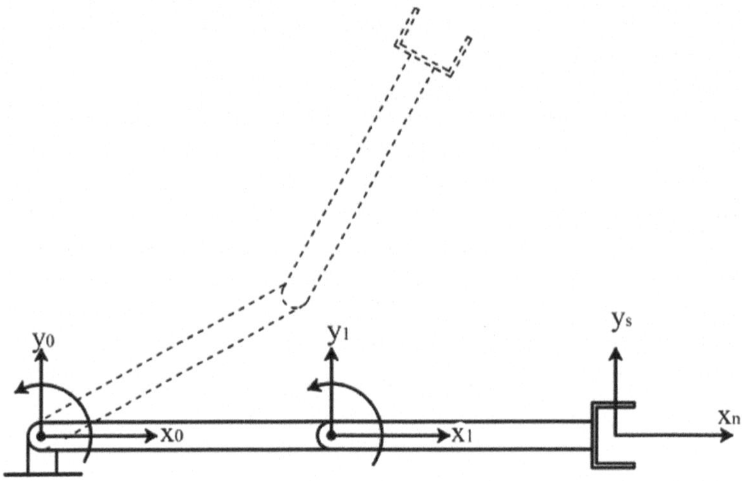

Fig. 4.7 Two-link planar manipulator

Example 4.1 (Two-link planar robot) Estimate the forward kinematics equations for the two-link planar robot manipulator shown in Fig. 4.7. Assume that the length of link 0 and link 1 is a_1 and a_2, respectively, where the second link starts from the point of connection with the other link to the centre of the end-effector.

Firstly, we have to assign the coordinate frames for this manipulator; the configuration in Fig. 4.7 (in a continuous line) will be chosen to be the rest configuration. The first joint, which is connected to the ground, is a revolute joint; therefore, z-direction will be in the direction of rotation. Assume the positive direction of the z-axis is out of the paper; then by the right-hand rule, x-axis and y-axis of the first coordinate can be determined. For this example, the thumb should be directed out of the paper to the reader while the four fingers directed along with link 1, and by using the explanation of the DH convention and right-hand rule, we can set coordinate frame $o_0x_0y_0z_0$ which is the base frame.

The four spatial parameters for link 0 can be identified between frame 1 and frame 2 while the other set of parameters for link 2 are specified between frame 2 and frame $(n\ s\ a)$. Referring to step 4 in the convention, it is clear that both decomposition angle and twist angle are zero, and Table 4.1 reveals the five parameters for the two links.

For simplicity, we will call frame $(n\ s\ a)$ Frame 3. Recalling Eq. (4.4), the homogeneous transformation matrices for the two links can be calculated as follows:

Table 4.1 Five spatial parameters for the two-link planar manipulator

Link ID	Frame	α	a	d	θ
1	$o_0x_0y_0z_0 \; - o_1x_1y_1z_1$	0	a_1	0	θ_1
2	$o_1x_1y_1z_1 \; - o_2x_2y_2z_2$	0	a_2	0	θ_2

$$H_1^0 = \begin{bmatrix} C_1 & -S_1 & 0 & a_1.C_1 \\ S_1 & C_1 & 0 & a_1.S_1 \\ 0 & 0 & 1 & 0 \\ 0 & 0 & 0 & 0 \end{bmatrix}, \tag{4.5}$$

$$H_2^1 = \begin{bmatrix} C_2 & -S_2 & 0 & a_2.C_2 \\ S_2 & C_2 & 0 & a_2.S_2 \\ 0 & 0 & 1 & 0 \\ 0 & 0 & 0 & 0 \end{bmatrix}. \tag{4.6}$$

The overall HTM is the multiplication of Eqs. (4.5) and (4.6) which is the forward kinematics equation for the planar robot

$$H_2^0 = H_1^0.H_2^1 = \begin{bmatrix} C_{12} & -S_{12} & 0 & a_1.C_1 + a_2.C_{12} \\ S_{12} & C_{12} & 0 & a_1.S_1 + a_2.S_{12} \\ 0 & 0 & 1 & 0 \\ 0 & 0 & 0 & 1 \end{bmatrix}. \tag{4.7}$$

Equation (4.7) shows that the position of the central point of the end-effector (P) relative to the base frame is

$$p_2^0 = \begin{bmatrix} x \\ y \\ z \end{bmatrix} = \begin{bmatrix} a_1.C_1 + a_2.C_{12} \\ a_1.S_1 + a_2.S_{12} \\ 0 \end{bmatrix}. \tag{4.8}$$

Also, the end-effector orientation relative to the base frame is

$$R_2^0 = \begin{bmatrix} C_{12} & -S_{12} & 0 \\ S_{12} & C_{12} & 0 \\ 0 & 0 & 1 \end{bmatrix}. \tag{4.9}$$

Example 4.2 (LabVolt 5150 manipulator) Figure 4.8 shows LabVolt 5150 manipulator (Ghafil and Jármai 2019), which is a 5DOF, four-link educational robot. The joint limits of the robot actuators, as well as link specifications, are given in Fig. 4.9. To start modelling the forward kinematics equations of this robot, zero-configuration was chosen to be as in Fig. 4.10. The same way as Example 4.1 has been followed to assign six-coordinate frames starting from the base point to the end-effector central point. Note that frame 3 and frame 4 coincide at the same point. Table 4.2 reveals the spatial parameters of the four links of the manipulator. Referring to Fig. 4.10, we have substituted the unknowns in Table 4.2 as follows:

$$d_1 = 255.55\,\text{mm}$$
$$a_2 = a_3 = 190\,\text{mm}$$

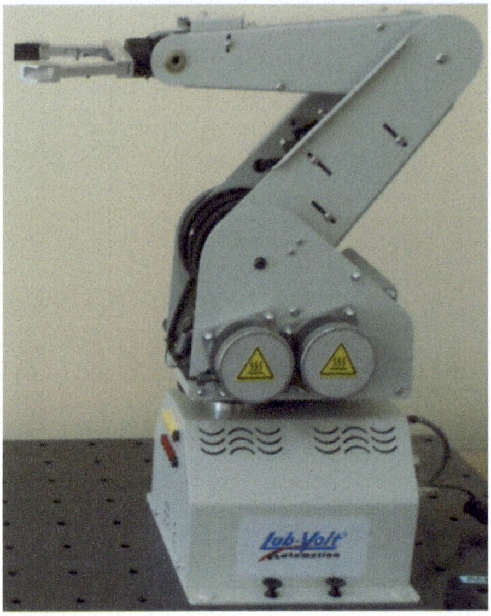

Fig. 4.8 LabVolt 5150 manipulator

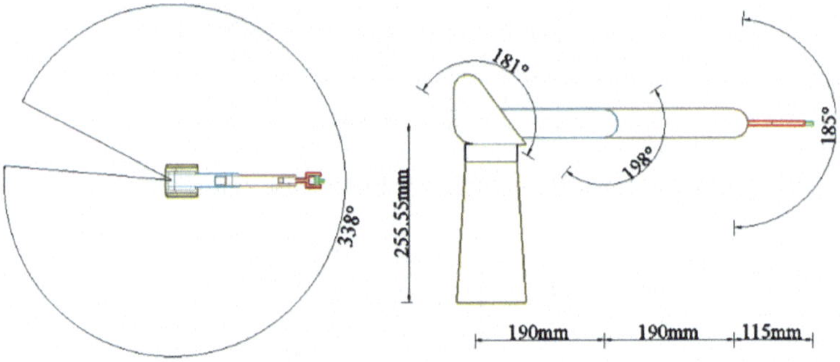

Fig. 4.9 Operative ranges and specifications of LabVolt 5150

$$d_5 = 115\,\text{mm}$$

The four HTMs can be determined by substituting each link parameter in Eq. (4.4). The set of HTMs will be

$$H_1^0 = \begin{bmatrix} C_1 & 0 & S_1 & 0 \\ S_1 & 0 & -C_1 & 0 \\ 0 & 1 & 0 & d_1 \\ 0 & 0 & 0 & 1 \end{bmatrix}, \tag{4.10}$$

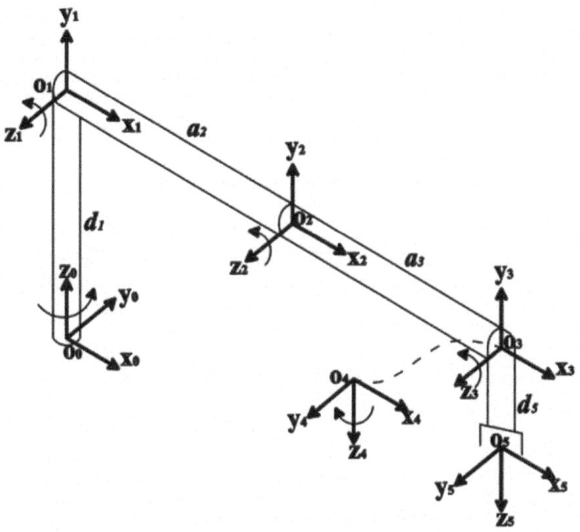

Fig. 4.10 Operative ranges and specifications of LabVolt 5150

Table 4.2 Spatial parameters of the LabVolt 5150 manipulator

Link ID	Frame	α	a	d	θ	Limit
1	$o_0x_0y_0z_0 - o_1x_1y_1z_1$	90	0	d_1	θ_1	−185, 153
2	$o_1x_1y_1z_1 - o_2x_2y_2z_2$	0	a_2	0	θ_2	−32, 149
3	$o_2x_2y_2z_2 - o_3x_3y_3z_3$	0	a_3	0	θ_3	−147, 51
4	$o_3x_3y_3z_3 - o_4x_4y_4z_4$	90	0	0	θ_4	−5180
5	$o_4x_4y_4z_4 - o_5x_5y_5z_5$	0	0	d_5	θ_5	−360, 360

$$H_2^1 = \begin{bmatrix} C_2 & -S_2 & 0 & a_2.C_2 \\ S_2 & C_2 & 0 & a_2.S_2 \\ 0 & 0 & 1 & 0 \\ 0 & 0 & 0 & 1 \end{bmatrix}, \tag{4.11}$$

$$H_3^2 = \begin{bmatrix} C_3 & -S_3 & 0 & a_3.C_3 \\ S_3 & C_3 & 0 & a_3.S_3 \\ 0 & 0 & 1 & 0 \\ 0 & 0 & 0 & 1 \end{bmatrix}, \tag{4.12}$$

$$H_4^3 = \begin{bmatrix} C_4 & 0 & S_4 & 0 \\ S_4 & 0 & -C_4 & 0 \\ 0 & 1 & 0 & 0 \\ 0 & 0 & 0 & 1 \end{bmatrix}, \tag{4.13}$$

$$H_5^4 = \begin{bmatrix} C_5 & -S_5 & 0 & 0 \\ S_5 & C_5 & 0 & 0 \\ 0 & 0 & 1 & d_5 \\ 0 & 0 & 0 & 1 \end{bmatrix}, \tag{4.14}$$

$$H_5^0 = H_1^0.H_2^1.H_3^2.H_4^3.H_5^4. \tag{4.15}$$

Equation (4.15) is 4×4 matrix holding the orientation and position vector of the end-effector with respect to the base frame, as revealed in Eqs. (4.16) and (4.17).

$$R_5^0 = \begin{bmatrix} S_1.S_5 + C_1.C_5.C_{234} & S_1.C_5 - C_1.S_5.C_{234} & C_1.S_{234} \\ -C_1.S_5 + S_1.C_5.C_{234} & C_1.S_5 + S_1.C_5.C_{234} & S_1.S_{234} \\ C_5.S_{234} & S_5.S_{234} & -C_{234} \end{bmatrix}, \tag{4.16}$$

$$P_5^0 = \begin{bmatrix} x = d_5 C_1 S_{234} + a_2 C_1 C_2 + a_3 C_1 C_{23} \\ y = d_5 S_1 S_{234} + a_2 C_2 S_1 + a_3 C_{23} \\ z = -d_5 C_{234} + a_2 S_2 + d_1 + a_3 S_{23} \end{bmatrix}. \tag{4.17}$$

The following code is written to calculate the forward kinematics equations for any set of joint angles.

```
clear;
clc;
format short G
t1 = input('Enter Theta 1:  ');       % joint angle 1
alpha_1 = 90;            %twist angle for link 1
a_1 = 0;                 %distance between z1 and z2
d_1 = 255.55;            % distance between x1 and x2
t2 = input('Enter Theta 2:  ');       % joint angle 2
alpha_2 = 0;             %twist angle for link 2
a_2 = 190;               %distance between z2 and z3
d_2 = 0;                 %distance between x2 and x3
t3 = input('Enter Theta 3:  ');       % joint angle 3
alpha_3 = 0;             %twist angle for link 3
a_3 = 190;               %distance between z3 and z4
d_3 = 0;                 % distance between x3 and x4
t4 = input('Enter Theta 4:  ');% joint angle 4
alpha_4 = 90;            %twist angle for link 4
a_4 = 0;                 %distance between z4 and z5
d_4 = 0;                 %distance between x4 and x5
t5 = input('Enter Theta 5:  ');       % joint angle 5
alpha_5 = 0;             %twist angle for link 5
```

```
a_5 = 0;                %distance between z5 and z6
d_5 = 115;              %distance between x5 and x6

H_1= HTM(t1,alpha_1,a_1,d_1); % HTM 1
H_2= HTM(t2,alpha_2,a_2,d_2); % HTM 2
H_3= HTM(t3,alpha_3,a_3,d_3); % HTM 3
H_4= HTM(t4,alpha_4,a_4,d_4); % HTM 4
H_5= HTM(t5,alpha_5,a_5,d_5); % HTM 5
TH = H_1*H_2*H_3*H_4*H_5                % over all HTM
```

The code for function HTM is

```
function [DH] = HTM(theta,alpha,a,d)
DH = [cosd(theta),-
1*sind(theta)*cosd(alpha),sind(theta)*sind(alpha),a*cos
d(theta);...
    sind(theta),cosd(theta)*cosd(alpha),-
1*cosd(theta)*sind(alpha),a*sind(theta);...
    0,sind(alpha),cosd(alpha),d;...
    0,0,0,1];
end
```

Example 4.3 (KUKA KR 100-2 PA robot) The KUKA KR 100-2 PA robot is a famous industrial manipulator with 6DOF revolute joints, shown in Fig. 4.11. The specifications of the robot were obtained from the datasheet of the robot online (Kuka_manual 2018).

The rest configuration was taken to be that in Fig. 4.12, which also explains the coordinate frames assignment according to the convention. For this assignment, the spatial parameters were concluded and given in Table 4.3.

$$H_1^0 = \begin{bmatrix} C_1 & 0 & -S_1 & a_1C_1 \\ S_1 & 0 & C_1 & a_1S_1 \\ 0 & -1 & 0 & 0 \\ 0 & 0 & 0 & 1 \end{bmatrix}, \tag{4.18}$$

$$H_2^1 = \begin{bmatrix} C_2 & -S_2 & 0 & a_2C_2 \\ S_2 & C_2 & 0 & a_2S_2 \\ 0 & 0 & 1 & 0 \\ 0 & 0 & 0 & 1 \end{bmatrix}, \tag{4.19}$$

$$H_3^2 = \begin{bmatrix} C_{3+90} & 0 & S_{3+90} & a_3 + C_{3+90} \\ S_{3+90} & 0 & -C_{3+90} & a_3 + S_{3+90} \\ 0 & 1 & 0 & 0 \\ 0 & 0 & 0 & 1 \end{bmatrix}, \tag{4.20}$$

Fig. 4.11 KUKA KR 100-2. Courtesy of KUKA robotics

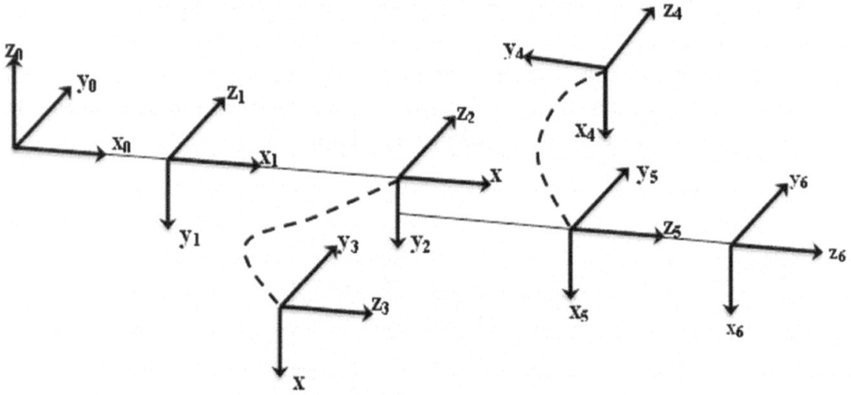

Fig. 4.12 KUKA KR 100-2 frame assignment

$$H_4^3 = \begin{bmatrix} C_4 & 0 & -S_4 & 0 \\ S_4 & 0 & C_4 & 0 \\ 0 & -1 & 0 & d_4 \\ 0 & 0 & 0 & 1 \end{bmatrix}, \tag{4.21}$$

Table 4.3 Spatial parameters of KUKA KR 100-2

Link ID	Frame	α	a	d	θ
1	$o_0x_0y_0z_0 - o_1x_1y_1z_1$	−90	350	0	θ_1
2	$o_1x_1y_1z_1 - o_2x_2y_2z_2$	0	1050	0	θ_2
3	$o_2x_2y_2z_2 - o_3x_3y_3z_3$	90	45	0	$\theta_3 + 90$
4	$o_3x_3y_3z_3 - o_4x_4y_4z_4$	−90	0	1000	θ_4
5	$o_4x_4y_4z_4 - o_5x_5y_5z_5$	90	0	0	θ_5
6	$o_5x_5y_5z_5 - o_6x_6y_6z_6$	0	0	210	θ_6

$$H_5^4 = \begin{bmatrix} C_5 & 0 & S_5 & 0 \\ S_5 & 0 & -C_5 & 0 \\ 0 & 1 & 0 & 0 \\ 0 & 0 & 0 & 1 \end{bmatrix}, \tag{4.22}$$

$$H_6^5 = \begin{bmatrix} C_6 & -S_6 & 0 & 0 \\ S_6 & C_6 & 0 & 0 \\ 0 & 0 & 1 & d_6 \\ 0 & 0 & 0 & 1 \end{bmatrix}. \tag{4.23}$$

Thus, the total HTM that relates the end-effector to the base frame will be

$$H_6^0 = H_1^0\, H_2^1\, H_3^2\, H_4^3\, H_5^4\, H_6^5. \tag{4.24}$$

As explained before, Eq. (4.24) has the following form:

$$H_6^0 = \begin{bmatrix} r_{11} & r_{12} & r_{13} & x \\ r_{21} & r_{22} & r_{23} & y \\ r_{31} & r_{32} & r_{33} & z \\ 0 & 0 & 0 & 1 \end{bmatrix}, \tag{4.25}$$

where rth elements are the components of the orientation matrix that represent the orientation of the end-effector with respect to the base frame, and x, y, and z coordinates are indexed as follows:

$$\begin{aligned} x &= T(1,4), \\ &= a_1C_1 + a_2C_1C_2 + a_3C_1C_{23+90} + d_4C_1S_{23+90} \\ &\quad - d_6\big[S_5(S_1S_4 - C_1C_4C_{23+90}) - C_1C_5S_{23+90}\big] \end{aligned} \tag{4.26}$$

$$y = T(2,4), \tag{4.27}$$

$$z = T(3,4), \tag{4.28}$$

The other components are left as a tutorial to the reader.

Example 4.4 (7DOF manipulator) In the following example, the more complicated manipulator shown in Fig. 4.13 has to be analysed for its forward kinematics. The process, as we have seen previously, is straightforward when frame assignment has been done as revealed in Fig. 4.13 using the DH convention; all of the parameters are explained in Table 4.4. It is worth mentioning that frames $o_5x_5y_5z_5$ and $o_6x_6y_6z_6$ are chosen to be located at the same point.

As mentioned before, the designer is free to choose where to put the origins of the coordinate frames as long as the convention is followed.

We have assumed that $a_2 = a_3 = a_4 = a_5 = a = 600$, $d_1 = 1000$, and $d_7 = 400$.

The homogeneous transformation matrices will be estimated using Eq. (4.4) and the parameters in Table 4.4 as follows:

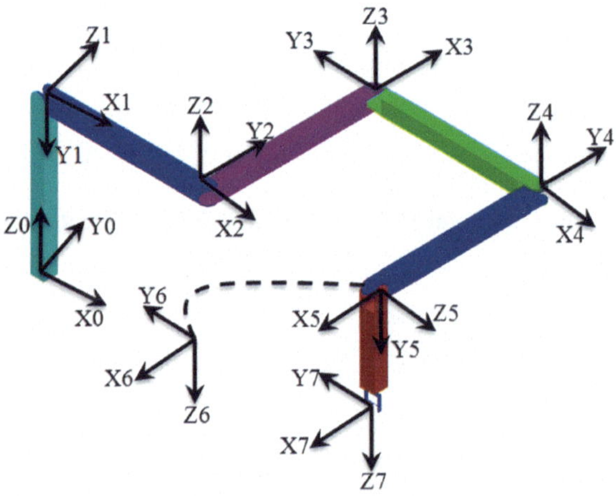

Fig. 4.13 Frame assignment of the 7DOF robot manipulator

Table 4.4 Spatial parameters of the 7DOF robot

Link ID	Frame	α	a	d	θ
1	$o_0x_0y_0z_0 - o_1x_1y_1z_1$	−90	0	d_1	θ_1
2	$o_1x_1y_1z_1 - o_2x_2y_2z_2$	90	a_2	0	θ_2
3	$o_2x_2y_2z_2 - o_3x_3y_3z_3$	0	a_3	0	$\theta_3 + 90$
4	$o_3x_3y_3z_3 - o_4x_4y_4z_4$	0	a_4	0	$\theta_4 - 90$
5	$o_4x_4y_4z_4 - o_5x_5y_5z_5$	−90	a_5	0	$\theta_5 - 90$
6	$o_5x_5y_5z_5 - o_6x_6y_6z_6$	−90	0	0	θ_6
7	$o_6x_6y_6z_6 - o_7x_7y_7z_7$	0	0	d_7	θ_7

$$H_1^0 = \begin{bmatrix} C_1 & 0 & -S_1 & 0 \\ S_1 & 0 & C_1 & 0 \\ 0 & -1 & 0 & d_1 \\ 0 & 0 & 0 & 1 \end{bmatrix}, \tag{4.29}$$

$$H_2^1 = \begin{bmatrix} C_2 & 0 & -S_2 & a\,C_2 \\ S_2 & 0 & -C_2 & a\,S_2 \\ 0 & 1 & 0 & 0 \\ 0 & 0 & 0 & 1 \end{bmatrix}, \tag{4.30}$$

$$H_3^2 = \begin{bmatrix} C_{3+90} & -S_{3+90} & 0 & a\,C_{3+90} \\ S_{3+90} & C_{3+90} & 0 & a\,S_{3+90} \\ 0 & 0 & 1 & 0 \\ 0 & 0 & 0 & 1 \end{bmatrix}, \tag{4.31}$$

$$H_4^3 = \begin{bmatrix} C_{4-90} & -S_{4-90} & 0 & a\,C_{4-90} \\ S_{4-90} & C_{4-90} & 0 & a\,S_{4-90} \\ 0 & 0 & 1 & 0 \\ 0 & 0 & 0 & 1 \end{bmatrix}, \tag{4.32}$$

$$H_5^4 = \begin{bmatrix} C_{5-90} & 0 & -S_{5-90} & a\,C_{5-90} \\ S_{5-90} & 0 & C_{5-90} & a\,S_{5-90} \\ 0 & -1 & 0 & 0 \\ 0 & 0 & 0 & 1 \end{bmatrix}, \tag{4.33}$$

$$H_6^5 = \begin{bmatrix} C_6 & 0 & -S_6 & 0 \\ S_6 & 0 & C_6 & 0 \\ 0 & -1 & 0 & 0 \\ 0 & 0 & 0 & 1 \end{bmatrix}, \tag{4.34}$$

$$H_7^6 = \begin{bmatrix} C_7 & -S_7 & 0 & 0 \\ S_7 & C_7 & 0 & 0 \\ 0 & 0 & 1 & d_7 \\ 0 & 0 & 0 & 1 \end{bmatrix}, \tag{4.35}$$

The total transformation matrix which relates the end-effector to the base frame will be

$$H_7^0 = H_1^0\, H_2^1\, H_3^2\, H_4^3\, H_5^4\, H_6^5\, H_7^6. \tag{4.36}$$

The following code was written to simulate the kinematic motion of this manipulator.

```
clear;
clc;
close all
format short G
th_1= input('Enter Theta 1:  ');      % joint angle 1
tw_1 = -90;        %twist angle for link 1
a_1 = 0;           %distance between z1 and z2
d_1 = 1000;        % distance between x1 and x2
th_2= input('Enter Theta 2:  ');      % joint angle 2
tw_2 = 90;        %twist angle for link 2
a_2 = 600;        %distance between z2 and z3
d_2 = 0;          % distance between x2 and x3
th_3= input('Enter Theta 3:  ')+90;  % joint angle 3
tw_3 = 0;         %twist angle for link 3
a_3 = 600;        %distance between z3 and z4
d_3 = 0;        % distance between x3 and x4
th_4= input('Enter Theta 4:  ')-90;  % joint angle 4
tw_4 = 0;         %twist angle for link 4
a_4 = 600;        %distance between z4 and z5
d_4 = 0;          % distance between x4 and x5
th_5= input('Enter Theta 5:  ')-90;     % joint angle 5
tw_5 = -90;       %twist angle for link 5
a_5 = 600;        %distance between z5 and z6
d_5 = 0;          % distance between x5 and x6
th_6= input('Enter Theta 6:  ');      % joint angle 6
tw_6 = -90;       %twist angle for link 6
a_6 = 0;          %distance between z6 and z7
d_6 = 0;          % distance between x6 and x7
th_7= input('Enter Theta 7:  ');      % joint angle 7
tw_7 = 0;         %twist angle for link 7
a_7 = 0;          %distance between z7 and zn
d_7 = 400;        % distance between x7 and xa

H_1= HTM(th_1,tw_1,a_1,d_1);
H_2= HTM(th_2,tw_2,a_2,d_2);
H_3= HTM(th_3,tw_3,a_3,d_3);
H_4= HTM(th_4,tw_4,a_4,d_4);
H_5= HTM(th_5,tw_5,a_5,d_5);
H_6= HTM(th_6,tw_6,a_6,d_6);
H_7= HTM(th_7,tw_7,a_7,d_7);
TH = H_1*H_2*H_3*H_4*H_5*H_6*H_7
cord_0 = zeros(4);
cord_1 =H_1;
cord_2 =H_1*H_2;
```

```
cord_3 =H_1*H_2*H_3;
cord_4 =H_1*H_2*H_3*H_4;
cord_5 =H_1*H_2*H_3*H_4*H_5;
cord_6 =H_1*H_2*H_3*H_4*H_5*H_6;
cord_7 =H_1*H_2*H_3*H_4*H_5*H_6*H_7;

drw_line(cord_0,cord_1);
hold on
grid on
drw_line(cord_1,cord_2);
drw_line(cord_2,cord_3);
drw_line(cord_3,cord_4);
drw_line(cord_4,cord_5);
drw_line(cord_5,cord_6);
drw_line(cord_6,cord_7);
axis([-2000 2500 -1000 1000 0 1500])
```

The code of the function HTM is the same as in Example 4.3, while the code for the function `drw_line` is:

```
function [] = drw_line(H_1,H_2)
plot3([H_1(1,4),H_2(1,4)],[H_1(2,4),H_2(2,4)],[H_1(3,4)
,H_2(3,4)],'k.-','MarkerSize',15);
end
```

Example 4.5 (Cylindrical robot with spherical wrist) This manipulator has 6DOF with two prismatic joints and three revolute joints. Figure 4.14 shows the robot, where the three last joints bounded by a dotted ellipse are called the spherical wrist.

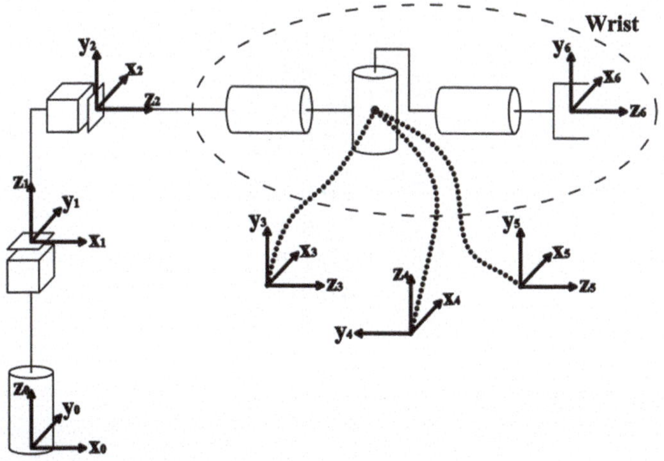

Fig. 4.14 Cylindrical robot with spherical wrist

Table 4.5 Link parameters of the cylindrical robot with spherical wrist

Link ID	Frame	α	a	d	θ
1	$o_0x_0y_0z_0 - o_1x_1y_1z_1$	0	0	d_1	θ_1
2	$o_1x_1y_1z_1 - o_2x_2y_2z_2$	90	0	P_2	0
3	$o_2x_2y_2z_2 - o_3x_3y_3z_3$	0	0	P_3	0
4	$o_3x_3y_3z_3 - o_4x_4y_4z_4$	−90	0	0	θ_4
5	$o_4x_4y_4z_4 - o_5x_5y_5z_5$	90	0	0	θ_5
6	$o_5x_5y_5z_5 - o_6x_6y_6z_6$	0	0	d_6	θ_6

The coordinate frames have been assigned using the convention as usual, where z-axis for each link is the direction of actuation. To simplify the calculations, we have chosen the origins of the last three joints of the wrist to coincide at the same point at the centre of the wrist. The simplification comes from dropping any additional offsets between coordinate frames. Equation (4.4) should be used for the revolute joint while Eq. (4.4a) is for prismatic joints.

Link parameters are shown in Table 4.5, where for link 2 and link 3 the parameter d is replaced by P (as we explained in Sect. 4.3) because these joints are prismatic and obey Eq. (4.4a).

Depending on the type of joint, links 1, 4, 5, and 6 are represented by Eq. (4.4) while links 2 and 3 are represented by Eq. (4.4a), and by considering Table 4.5 parameters, we can write link equations as follows:

$$H_1^0 = \begin{bmatrix} C_1 & -S_1 & 0 & 0 \\ S_1 & C_1 & 0 & 0 \\ 0 & 0 & 1 & d_1 \\ 0 & 0 & 0 & 1 \end{bmatrix}, \tag{4.37}$$

$$H_2^1 = \begin{bmatrix} 0 & 0 & 1 & 0 \\ 1 & 0 & 0 & 0 \\ 0 & 1 & 0 & P_2 \\ 0 & 0 & 0 & 1 \end{bmatrix}, \tag{4.38}$$

$$H_3^2 = \begin{bmatrix} 1 & 0 & 0 & 0 \\ 0 & 1 & 0 & 0 \\ 0 & 0 & 1 & P_3 \\ 0 & 0 & 0 & 1 \end{bmatrix}, \tag{4.39}$$

$$H_4^3 = \begin{bmatrix} C_4 & 0 & -S_4 & 0 \\ S_4 & 0 & C_4 & 0 \\ 0 & -1 & 0 & 0 \\ 0 & 0 & 0 & 1 \end{bmatrix}, \tag{4.40}$$

$$H_5^4 = \begin{bmatrix} C_5 & 0 & S_5 & 0 \\ S_5 & 0 & -C_5 & 0 \\ 0 & 1 & 0 & 0 \\ 0 & 0 & 0 & 1 \end{bmatrix}, \tag{4.41}$$

$$H_6^5 = \begin{bmatrix} C_1 & -S_1 & 0 & 0 \\ S_1 & C_1 & 0 & 0 \\ 0 & 0 & 1 & d_6 \\ 0 & 0 & 0 & 1 \end{bmatrix}, \tag{4.42}$$

The following code is for simulating the motion of this type of manipulator:

```
clear;
clc;
close all
format short G
th_1= input('Enter Theta 1 in deg:  ');      % joint
angle 1
tw_1 = 0;            %twist angle for link 1
a_1 = 0;             %distance between z1 and z2
d_1 = 400;           % distance between x1 and x2
P_2= input('Enter  p2 in mm:  ');% joint displacement 2
tw_2 = 90;           %twist angle for link 2
a_2 = 0;             %distance between z2 and z3
P_3 = input('Enter  p3 in mm:  ');  % joint angle 3
tw_3 = 0;            %twist angle for link 3
a_3 = 0;             %distance between z3 and z4
th_4= input('Enter  Theta 4 in deg:  ');  % joint angle
4
tw_4 = -90;          %twist angle for link 4
a_4 = 0;             %distance between z4 and z5
d_4 = 0;             % distance between x4 and x5
th_5= input('Enter  Theta 5 in deg:  ');      % joint
angle 5
```

```
tw_5 = 90;          %twist angle for link 5
a_5 = 0;            %distance between z5 and z6
d_5 = 0;            % distance between x5 and x6
th_6= input('Enter Theta 6 in deg:  ');      % joint
angle 6
tw_6 = 0;           %twist angle for link 6
a_6 = 0;            %distance between z6 and z7
d_6 = 500;          % distance between x6 and x7
H_1= HTM(th_1,tw_1,a_1,d_1)
H_2= HTM_P(0,tw_2,a_2,P_2)
H_3= HTM_P(0,tw_3,a_3,P_3)
H_4= HTM(th_4,phai_4,tw_4,a_4,d_4)
H_5= HTM(th_5,phai_5,tw_5,a_5,d_5)
H_6= HTM(th_6,phai_6,tw_6,a_6,d_6)
TH = H_1*H_2*H_3*H_4*H_5*H_6
cord_0 = zeros(4);
cord_1 =H_1;
cord_2 =H_1*H_2;
cord_3 =H_1*H_2*H_3;
cord_4 =H_1*H_2*H_3*H_4;
cord_5 =H_1*H_2*H_3*H_4*H_5;
cord_6 =H_1*H_2*H_3*H_4*H_5*H_6;
drw_line(cord_0,cord_1);
hold on
grid on
drw_line(cord_1,cord_2);
drw_line(cord_2,cord_3);
drw_line(cord_3,cord_4);
drw_line(cord_4,cord_5);
drw_line(cord_5,cord_6);
axis([-2000 2500 -1000 1000 0 1500])
```

The code of `drw_line` and `HTM` functions is the same as those mentioned before while the code for `HTM_P` function is:

```
function [DH] = Denavit(theta,alpha,a,d)
DH = [cosd(theta),-
1*sind(theta)*cosd(alpha),sind(theta)*sind(alpha),a*cos
d(theta);...
    sind(theta),cosd(theta)*cosd(alpha),-
1*cosd(theta)*sind(alpha),a*sind(theta);...
    0,sind(alpha),cosd(alpha),d;...
    0,0,0,1];
end
```

4.5 Inverse Kinematics

This is the problem of finding the joint variables from the given position and orientation of the end-effector. While forward kinematics detects the position and orientation of the end-effector from the given set of joint variables, inverse kinematics is the inverse operation, but it is somewhat complicated.

$$\begin{bmatrix} \theta_1 \\ \theta_2 \\ \vdots \\ \theta_n \end{bmatrix} \begin{array}{c} \Rightarrow \\ \Leftarrow \end{array} \begin{bmatrix} r_{11} & r_{12} & r_{13} & x \\ r_{21} & r_{22} & r_{23} & y \\ r_{31} & r_{32} & r_{33} & z \\ 0 & 0 & 0 & 1 \end{bmatrix}. \tag{4.43}$$

There are many approaches mentioned in other books to deal with this problem, like closed-form and a geometric approach. These methods are robot-dependent and differ from one manipulator to another. It is possible to find the solution for a 6DOF manipulator but hardly possible—or indeed impossible—to find a theoretical solution for a different 5DOF. In this book, we will only focus on the solution from the viewpoint of optimization algorithms in order to develop a unified procedure for any robot manipulator. Since forward equations are a straightforward process, we will rely on these equations to establish the objective function for the inverse problem. Remember back to Sect. 2.21, where we implemented a particle swarm optimization algorithm and artificial bee colony algorithm. The objective function example of PSO was the sphere function, and the algorithm has to find the best set of variables that satisfy the global minimum of the sphere function. Here, we are also looking for an optimum set of joint variables that can lead to the minimum of a cost function, so the only thing to do is develop a cost function for the inverse kinematics.

Consider Fig. 4.15, where for a specific robot configuration, the current position vector of the end-effector can be represented by the distance from the base of the end-effector of the manipulator while the desired position vector represents the task point. Obviously, if the difference between these two vectors is zero, then the tooltip

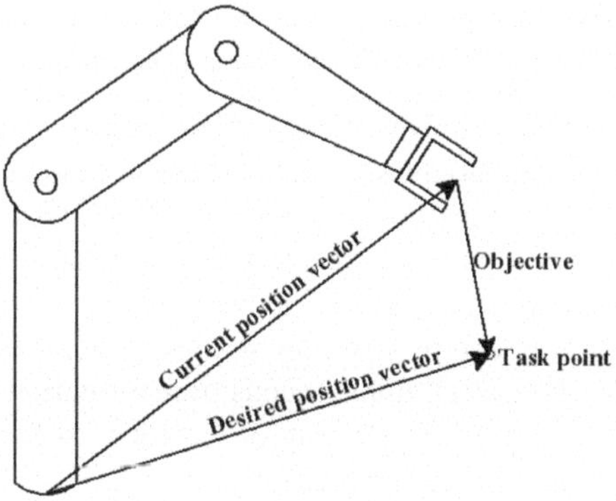

Fig. 4.15 Representation of the objective function for the inverse kinematics problem

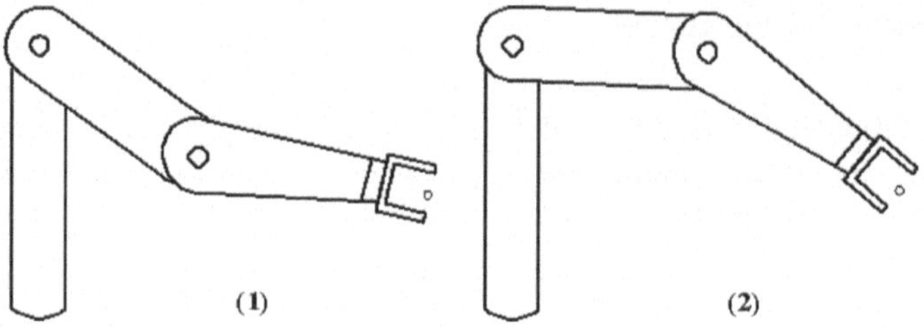

Fig. 4.16 Different orientations for the same position

will be in the right position at the task point, and this is the objective function f of the inverse problem

$$f = \|\mathrm{Ci} - \mathrm{De}\|, \tag{4.44}$$

where Ci denotes the instantaneous position vector and De is the desired position vector. In other words, Eq. (4.44) is the function that has to be minimized as much as possible, and it is just the distance between the end-effector and task point.

$$f = \sqrt{(x_{\mathrm{Ci}} - x_t)^2 + (y_{\mathrm{Ci}} - y_t)^2 + (z_{\mathrm{Ci}} - z_t)^2}. \tag{4.45}$$

where t refers to the task point coordinates given for the inverse kinematics problem.

If Eq. (4.45) has been used alone as an objective function, we may get the end-effector in the task point but with many choices of orientations. Figure 4.16 reveals an example of two options of orientation for the same task point, and in manipulators with a high degree of freedom, there can be even an infinite number of orientations.

4.5.1 Objective Function

In this section, we shall model the objective function for the inverse kinematics of any robot manipulator. Figure 4.17 shows a schematic diagram for the inverse problem; it is more descriptive to explain the procedure by a set of notes as follows:

1. The optimization algorithm sends the candidate solution, which is a set of possible joint variables, to the cost function to evaluate its fitness.
2. Cost function contains the desired task point coordinates; it sends the possible solution to forward function to get x, y, and z coordinates of the tooltip.
3. Forward function contains all the forward kinematics equations of the robot arm, and by substituting the candidate solution into those equations, we can get the overall homogenous transformation matrix by a repeated call for the HTM function. The output of the forward function is the position vector of the total transformation matrix.

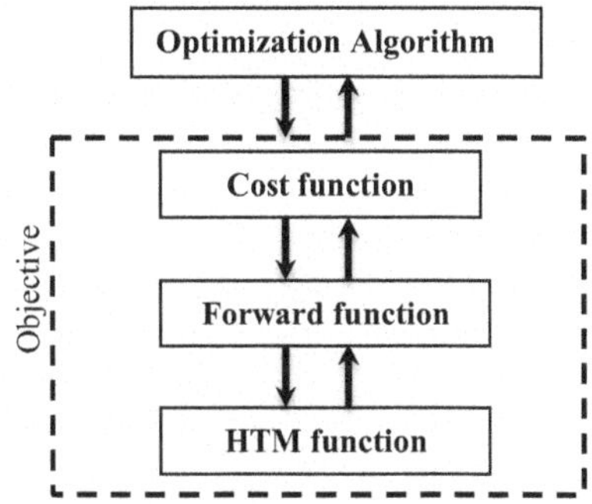

Fig. 4.17 Objective function scheme for inverse kinematics problem

4. Cost function will receive the position vector and apply Eq. (4.45) to the candidate vector and the desired task position vector. The result is the fitness of the solution, which will be returned to the main optimization algorithm.

This process is constant for all manipulators; the only thing to change is the forward kinematics equations and the task position vector. It is worth mentioning that not all optimization methods can return a guaranteed solution; this depends on the efficiency of the algorithm itself.

Example 4.6 (Inverse Kinematic of LabVolt 5150) In Example 4.2, we have formulated the forward kinematics equations for LabVolt 5150 manipulator from Eqs. (4.10) to (4.15). Now, we want to find the appropriate set of joint angles corresponding to the end-effector position vector of $(495, 0, 255.55)^T$. By using the procedure in Sect. 4.5.1, which is represented by Fig. 4.17, we have modelled the objective function of the robot. In the following, we have used the ABC algorithm from Sect. 2.21.2 to find the best joint angles corresponding to the given position vector.

ABC algorithm:

```
clc;
clear;
close all;
%% Problem Definition
CostFunction=@(x) cost(x);
nVar=5;
VarSize=[1 nVar];
```

The range of variables is the limit of joint angles of the robot given in Table 4.2.

```
VarMin= [-185,-32,-147,-5,-360];    % Decision Variables
Lower Bound
VarMax= [153,149,51,180,360];       % Decision Variables
Upper Bound
%% ABC parameters
nPop=500;                       % Population Size
MaxIt=1000;                      % Maximum Number of Iterations
employed = nPop/2;              % Number of employed Bees
nOnlooker=nPop/2;               % Number of Onlooker Bees
Limit = nOnlooker*nPop;             % Limit parameter
%% Initialization
templete_bee.Position=[];
templete_bee.Cost=[];
Bee = repmat(templete_bee,nPop,1);
BestSol.Cost=inf;
for i=1:nPop
    % scout bees phase
    for j=1:length(VarMin)
      Bee(i).Position(j) =
unifrnd(VarMin(j),VarMax(j),1) ;
    end
    Bee(i).Cost=CostFunction(Bee(i).Position);
    if Bee(i).Cost<=BestSol.Cost
        BestSol=Bee(i);
    end
end
% Abandonment Counter
C=zeros(nPop,1);
% Array to Hold Best Cost Values
BestCost=zeros(MaxIt,1);

%% ABC Main Loop
for it=1:MaxIt
    % Employed Bees
    for i=1:employed
        phi=unifrnd(-1,+1,VarSize);
        k = randi(employed);
        % New Bee Position

newbee.Position=Bee(i).Position+phi.*(Bee(i).Position-
Bee(k).Position);
        % limits
        for j=1:length(VarMin)
            newbee.Position(j) =
```

```
max(newbee.Position(j), VarMin(j));
            newbee.Position(j) =
min(newbee.Position(j), VarMax(j));
        end
        % Evaluation
        newbee.Cost=CostFunction(newbee.Position);
        % Comparision
        if newbee.Cost < Bee(i).Cost
            Bee(i)=newbee;
        else
            C(i)=C(i)+1;
        end
    end
F=zeros(nPop,1);
for a=1: nPop
    if(Bee(a).Cost>=0)
        F(a)= 1/(1 + Bee(a).Cost);
    else
        F(a) = 1+ abs(Bee(a).Cost);
    end
end
P = F/sum(F);
    % Onlooker Bees
    for m=1:nOnlooker
        % Select Source Site
        i=RouletteWheelSelection(P);
        k = randi(nPop);
        phi=unifrnd(-1,+1,VarSize);
        % New Bee Position

newbee.Position=Bee(i).Position+phi.*(Bee(i).Position-
Bee(k).Position);
        % limits
        for j=1:length(VarMin)
            newbee.Position(j) =
max(newbee.Position(j), VarMin(j));
            newbee.Position(j) =
min(newbee.Position(j), VarMax(j));
        end
        % Evaluation
        newbee.Cost=CostFunction(newbee.Position);
        % Comparison
        if newbee.Cost<=Bee(i).Cost
            Bee(i)=newbee;
        else
            C(i)=C(i)+1;
        end
```

```
    end
    % Scout Bees
    for i=1:nPop
        if C(i)>=Limit
            for j=1:length(VarMin)
              Bee(i).Position(j) =
unifrnd(VarMin(j),VarMax(j),1) ;
            end
            % limits
                for j=1:length(VarMin)
                  newbee.Position(j) =
max(newbee.Position, VarMin(j));
                  newbee.Position(j) =
min(newbee.Position, VarMax(j));
                end
            Bee(i).Cost=CostFunction(Bee(i).Position);
            C(i)=0;
        end
    end
    % Update Best Solution Ever Found
    for i=1:nPop
        if Bee(i).Cost<=BestSol.Cost
            BestSol=Bee(i);
        end
    end
    % Store Best Cost Ever Found
    BestCost(it)=BestSol.Cost;
    % Display Iteration Information
    disp(['Iteration ' num2str(it) ': Best Cost = '
num2str(BestCost(it))]);
end
%% Results
figure;
%plot(BestCost,'LineWidth',2);
semilogy(BestCost,'LineWidth',2);
xlabel('Iteration');
ylabel('Best Cost');
grid on;
```

Cost function:

```
function obj = cost(sol)
t1 = sol(1);
t2 = sol(2);
t3 = sol(3);
t4= sol(4);
t5= sol(5);
[x,y,z] = Forward(t1,t2,t3,t4,t5);
v = [495,0,255.55];
obj = sqrt((x-v(1))^2+(y-v(2))^2+(z-v(3))^2);
end
```

Forward function

```
function [x,y,z] = Forward(t1,t2,t3,t4,t5)
alpha_1 = 90;           %twist angle for link 1
a_1 = 0;                %distance between z1 and z2
d_1 = 255.55;           % distance between x1 and x2
alpha_2 = 0;            %twist angle for link 2
a_2 = 190;              %distance between z2 and z3
d_2 = 0;                %distance between x2 and x3
alpha_3 = 0;            %twist angle for link 3
a_3 = 190;              %distance between z3 and z4
d_3 = 0;                % distance between x3 and x4
alpha_4 = 90;           %twist angle for link 4
a_4 = 0;                %distance between z4 and z5
d_4 = 0;                % distance between x4 and x5
alpha_5 = 0;           %twist angle for link 5
a_5 = 0;               %distance between z5 and z6
d_5 = 115;             %distance between x5 and x6

H_1= HTM(t1,alpha_1,a_1,d_1); % HTM 1
H_2= HTM(t2,alpha_2,a_2,d_2); % HTM 2
H_3= HTM(t3,alpha_3,a_3,d_3); % HTM 3
H_4= HTM(t4,alpha_4,a_4,d_4); % HTM 4
H_5= HTM(t5,alpha_5,a_5,d_5); % HTM 5
TH = H_1*H_2*H_3*H_4*H_5;                 % over all HTM
x=TH(1,4);
y=TH(2,4);
z=TH(3,4);
end
```

The HTM function was mentioned in Example 4.5.

As we have mentioned before, the fact that the modelling objective function depends only on the difference between the current position vector and task point vector may lead to many or infinite possible sets of joint angles that correspond to the given point in the Cartesian space. In some applications where there is a specific orientation for the position of the end-effector, we should add the differences of orientation matrix elements of the candidate and the desired solution to Eq. (4.45) to estimate the cost of the solution.

$$f = \sqrt{(x_{\mathrm{Ci}} - x_{\mathrm{t}})^2 + (y_{\mathrm{Ci}} - y_{\mathrm{t}})^2 + (z_{\mathrm{Ci}} - z_{\mathrm{t}})^2} + (\mathrm{Rc}_{i,j} - \mathrm{Rt}_{i,j}), \tag{4.46}$$

where Rc and Rt refer to candidate and task rotation matrices, respectively, and i, $j = 1, 2, 3$. If there is a set of Euler angles or roll, pitch, and yaw angles for the end-effector, we can use Eqs. (3.22) or (3.23) to find the orientation matrix of the tooltip.

One can recognize that inverse kinematic solution is easily computed by optimization techniques. However, for a high degree of freedom the solution of the optimization algorithm required moderated time and maybe not appropriate for real-time applications that need an instantaneous response. Another quick alternative solution can be provided for such applications using artificial neural network (Ghafil et al. 2019).

4.6 Summary

Homogenous transformation matrix driven in chapter three is employed by Denavit–Hartenberg convention to formulate forward kinematic equations. Forward kinematics is the problem of finding the position and orientation of the end-effector given joint variables. Inverse kinematics is finding the set of joint variables given end-effector position and orientation. Optimization algorithms use a set of joint variables as optimization vector. The optimum set of joint variables is corresponding to the minimum distance between the position of the end-effector and the desired position in Cartesian space. This minimum distance is the objective function of the inverse kinematic problem.

Exercises

4.1 For the 11DOF manipulator shown in Fig. 4.18:

   a. Formulate the forward kinematics equations.
   b. Consider different coordinate frame assignments where z_2, z_3 should be in the opposite direction.

4.2 Consider the LabVolt manipulator in Example 4.6. Find the joint angles that correspond to the end-effector position and orientation represented by the following HTM. Include orientation in your objective function.

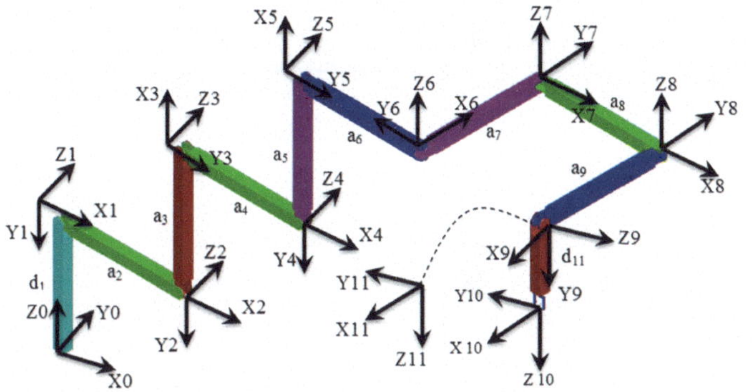

Fig. 4.18 An 11-degrees-of-freedom robot

$$H = \begin{bmatrix} 0.1112 & -0.8074 & 0.5795 & 336.3267 \\ -0.9358 & 0.1112 & 0.3346 & 194.1783 \\ -0.3346 & -0.5795 & -0.7431 & 332.1964 \\ 0 & 0 & 0 & 1 \end{bmatrix}$$

4.3 Figure 4.19 shows two revolute joints and one prismatic joint manipulator RRP. Formulate the forward kinematics equation, and solve the inverse problem using one of the optimization algorithms in Chap. 2.

4.4 Apply the same requirements as in Exercise 3 to the one revolute joint, one prismatic joint, and one revolute joint RPR manipulator in Fig. 4.20.

4.5 Figure 4.21 shows a PUMA robot, which is a 6DOF manipulator with a spherical wrist. Follow a similar procedure to Example 4.5 to analyse the robot manipulator.

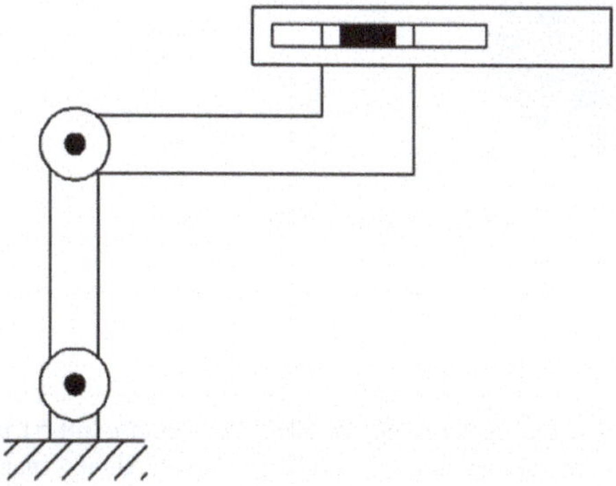

Fig. 4.19 RRP planar manipulator

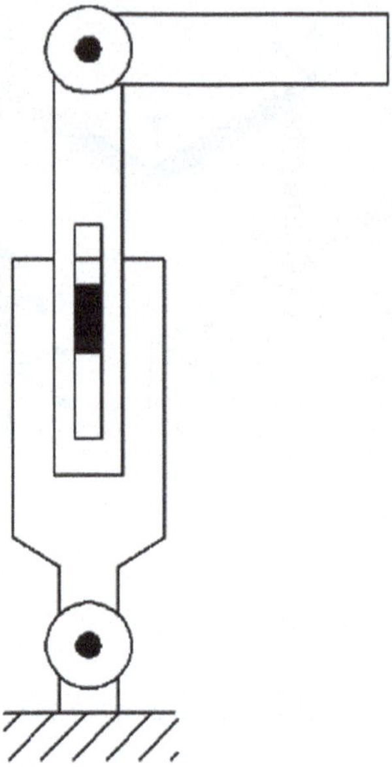

Fig. 4.20 RPR planar manipulator

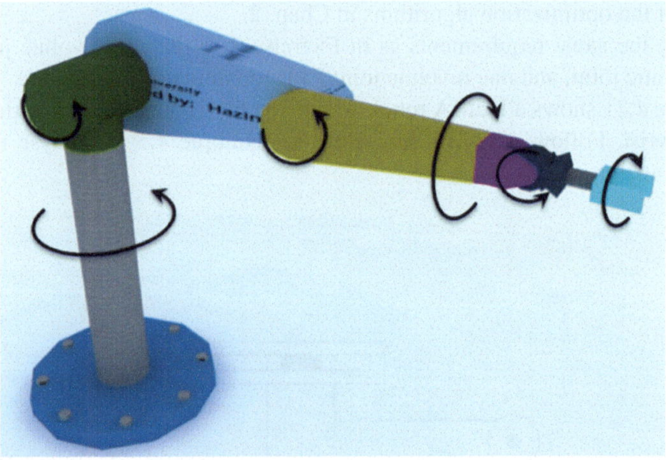

Fig. 4.21 6DOF revolute joints (PUMA) robot manipulator

References

Ghafil HN, Jármai K (2019) On performance of ED7500 and lab-volt 5150 5DOF manipulators: calibration and analysis. In: MultiScience—XXXIII. MicroCAD international multidisciplinary scientific conference University of Miskolc, 23–24 May 2019. ISBN 978-963-358-177-3. https://doi.org/10.26649/musci.2019.042

Ghafil HN, Kovács L, Jármai K (2019) Investigating three learning algorithms of a neural networks during inverse kinematics of robots. In: Solutions for sustainable development: proceedings of the 1st international conference on engineering solutions for sustainable development (ICESSD 2019), 3–4 Oct 2019. CRC Press, Miskolc, Hungary, 2019, pp 33–40

Kuka_manual (2018) KUKA. https://www.google.hu/url?sa=t&rct=j&q=&esrc=s&source=web&cd=1&ved=0ahUKEwjuurbM6-jYAhXSZVAKHUoCC6kQFggoMAA&url=https%3A%2F%2Fwww.kuka.com%2F-%2Fmedia%2Fkuka-downloads%2Fimported%2F6b77eecacfe542d3b736af377562ecaa%2Fdb_kr_100_2_comp_de.pdf&usg=AOvVaw1uAaR6NX0xxacpztuvEL0n. Accessed 14 Oct 2018

Chapter 5
The Manipulator Jacobian

5.1 Introduction

In this chapter, we will consider the representation of the velocities and accelerations of points on the manipulator. Velocity kinematics becomes relevant when we speak about industrial robot manipulators because of its importance in the manufacturing process. For instance, we have to consider velocity kinematics when we design the controller of manipulators for welding, cutting, painting, grinding, etc. Furthermore, in assembly processes in heavy industry, we have to be more careful about the acceleration of the robot actuators because it is a crucial factor in the failure and lifetime of the robots.

5.2 Velocity of a Point

Consider Fig. 5.1, when a rigid body moves in pure rotation all the points on the body $P1, P2, \ldots Pn$. The points sweep out in a circle and their centres are the axis of rotation of the body; it passes through the central point in Fig. 5.1. These circles may have the same or different radii, which are perpendicular from a point to the axis of rotation. During rotation, each radius sweeps out at the same angular displacement θ while moving from P to $\overline{P}$, as shown in Fig. 5.2. Thus, the angular velocity w is the time rate of changing the angular displacement and by assuming that k is a unit vector in the direction of rotation, we can say that

$$w = \frac{\mathrm{d}\theta}{\mathrm{d}t}k = \dot{\theta}\,k. \tag{5.1}$$

While every point on the body is subjected to the same angular velocity, each point has individual linear velocities v depending on their radii

H. N. Ghafil and K. Jármai (ed.), *Optimization for Robot Modelling with MATLAB*,
https://doi.org/10.1007/978-3-030-40410-9_5

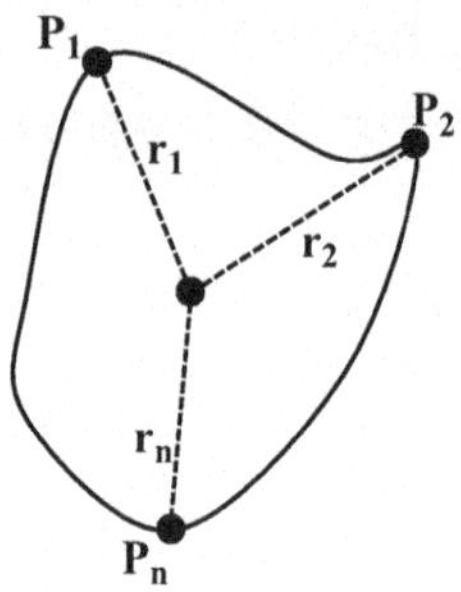

Fig. 5.1 Different points on a body sweep in different circles on the same axis of rotation

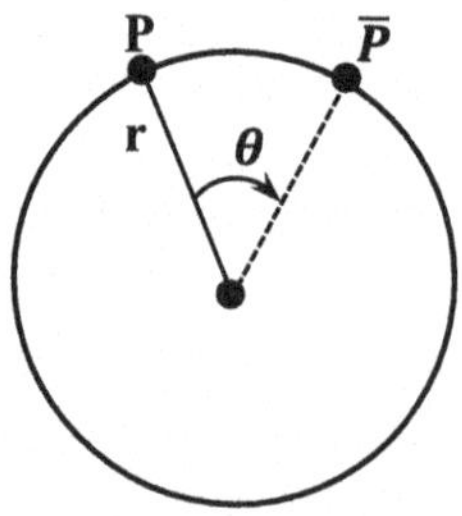

Fig. 5.2 Point sweeps out in a circle when tracing angular displacement

$$v = w \times r. \tag{5.2}$$

For the purposes of our work, we are interested in describing the angular velocity of a coordinate frame and the linear velocity of a point associated with that frame relative to some reference coordinate frame.

5.3 Skew-Symmetric Matrix

In this section, we will introduce the notion of the skew-symmetric matrix, which is useful for representation of the velocities in general form. Any matrix is said to be skew-symmetric S if and only if

$$S + S^T = 0 \quad i, j = 1, 2, \ldots, n. \tag{5.3}$$

We can write Eq. (5.3) in terms of the component of the matrix S using tensor notation

$$S_{i,j} + S_{j,i} = 0. \tag{5.4}$$

The second term in Eq. (5.4) is the transpose of the matrix S, which is simply done by flipping the order of the i and j components. Equation (5.4) clearly shows that the diagonal of the skew-symmetric matrix, where $i = j$, has zero values. Also, for all $i \neq j$ the off-diagonal satisfies $S_{i,j} = -S_{j,i}$. For instance, if $i = j = 1$ then

$$S_{11} + S_{11} = 0 \Rightarrow S_{11} = 0. \tag{5.5}$$

If $i = 1$ and $j = 2$

$$S_{12} + S_{21} = 0 \Rightarrow S_{12} = -S_{21}. \tag{5.6}$$

For every 3×3 symmetric matrix, there should be only three entries that form the matrix

$$S = \begin{bmatrix} 0 & -S_3 & S_2 \\ S_3 & 0 & -S_1 \\ -S_2 & S_1 & 0 \end{bmatrix}. \tag{5.7}$$

Example 5.1 If we have the vector $a = \begin{bmatrix} 1 & 2 & 3 \end{bmatrix}$, then the skew-symmetric matrix of the vector should be

$$S(a) = \begin{bmatrix} 0 & -3 & 2 \\ 3 & 0 & -1 \\ -2 & 1 & 0 \end{bmatrix}. \tag{5.8}$$

Also, for the three basic unit vectors $i = \begin{bmatrix} 1 \\ 0 \\ 0 \end{bmatrix}$, $j = \begin{bmatrix} 0 \\ 1 \\ 0 \end{bmatrix}$, and $k = \begin{bmatrix} 0 \\ 0 \\ 1 \end{bmatrix}$ the skew-symmetric matrices are given as

$$S(i) = \begin{bmatrix} 0 & 0 & 0 \\ 0 & 0 & -1 \\ 0 & 1 & 0 \end{bmatrix}, \; S(j) = \begin{bmatrix} 0 & 0 & 1 \\ 0 & 0 & 0 \\ -1 & 0 & 0 \end{bmatrix}, \text{ and } S(j) = \begin{bmatrix} 0 & 0 & 1 \\ 0 & 0 & 0 \\ -1 & 0 & 0 \end{bmatrix} \tag{5.9}$$

5.3.1 Properties of Skew-Symmetric Matrix

Skew-symmetric matrix of a vector has the following properties

1. For vectors a and b as well as scalar values α and β, the operator S is linear.

$$S(\alpha a + \beta b) = aS(\alpha) + bS(\beta)$$

2. For all $a, b \in \mathbb{R}^3$

$$S(a)b = a \times b \tag{5.10}$$

3. For all $a, b \in \mathbb{R}^3$ and $R \in SO(3)$

$$R(a \times b) = Ra \times Ra. \tag{5.11}$$

4. For all $a \in \mathbb{R}^3$ and $R \in SO(3)$

$$R\,S(a)\,R^T = S(Ra). \tag{5.12}$$

5. For all $a \in \mathbb{R}^3$

$$S(a) + (S(a))^T = 0. \tag{5.13}$$

5.4 Angular Velocity

For all $\theta \in \mathbb{R}$ and orthogonal rotational matrix $R \in SO(3)$

$$R(\theta)R(\theta)^T = I. \tag{5.14}$$

By taking the derivative of both sides of Eq. (5.14), we get

$$\frac{\mathrm{d}R(\theta)}{\mathrm{d}\theta}R(\theta)^T + R(\theta)\frac{\mathrm{d}R(\theta)^T}{\mathrm{d}\theta} = 0. \tag{5.15}$$

Now, define

$$s := \frac{\mathrm{d}R(\theta)}{\mathrm{d}\theta}R(\theta)^T, \tag{5.16}$$

$$s^T = R(\theta)\frac{\mathrm{d}R(\theta)^T}{\mathrm{d}\theta}. \tag{5.17}$$

Adding Eqs. (5.12) and (5.13) yields the same results in Eq. (5.11); this means that s is a skew-symmetric matrix according to the property in (5.9). Multiplying both sides of Eq. (5.12) by $R(\theta)$ from the right direction

$$S\,R(\theta) = \frac{\mathrm{d}R(\theta)}{\mathrm{d}\theta}\underbrace{R(\theta)^T R(\theta)}_{I} = \frac{\mathrm{d}R(\theta)}{\mathrm{d}\theta}. \tag{5.18}$$

Equation (5.14) says that the derivative of the rotation matrix is only the multiplication of the skew-symmetric and the rotation matrices

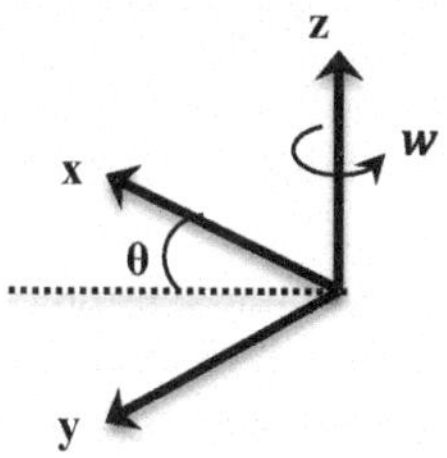

Fig. 5.3 Frame rotated about its *z*-axis

$$\frac{dR(\theta)}{d\theta} = S\,R(\theta). \tag{5.19}$$

If the rotation matrix is a function of time in case of rotational angle varying with time $R = R(t)$, then the following equation is identical to Eq. (5.19)

$$\dot{R}(t) = S(t)R(t). \tag{5.20}$$

Equation (5.16) is valid under the assumption that R is differentiable for all $t \in \mathbb{R}$. By using the chain rule, we can calculate the time derivative of R.

$$\dot{R} = \frac{dR}{dt} = \frac{dR}{d\theta}\frac{d\theta}{dt} = \dot{\theta}\frac{dR}{d\theta}. \tag{5.21}$$

Using the identical Eq. (5.16), we can say that

$$\dot{R}(t) = \dot{\theta}\,S\,R(t). \tag{5.22}$$

Equation (5.18) is a general equation to find the time derivative of a rotation matrix for a coordinate frame rotating about an arbitrary axis. Consider Fig. 5.3, where the coordinate frame rotates about the *z*-axis and, consequently, the angular velocity in the direction of the unit vector k. In this case, we can find the time derivative of the rotational matrix by re-writing Eq. (5.18) as follows

$$\dot{R}(t) = \dot{\theta}\,S(k)\,R(t) = S(w(t))R(t). \tag{5.23}$$

where $w(t) = \dot{\theta}k$ is the angular velocity in the direction of $k = \begin{bmatrix} 0 & 0 & 1 \end{bmatrix}^T$. Thus, Eq. (5.18) calculates the time derivative of the rotational matrix of a possibly moving frame and swaps angular displacement in the direction of an arbitrary unit vector. In other words, given the scalar value of the angular velocity in the direction of $i = \begin{bmatrix} 1 & 0 & 0 \end{bmatrix}^T$, $j = \begin{bmatrix} 0 & 1 & 0 \end{bmatrix}^T$, or $k = \begin{bmatrix} 0 & 0 & 1 \end{bmatrix}^T$ we can find the proper skew-symmetric matrix and multiply it by the rotational matrix to find $\dot{R}(t)$.

5.5 Angular Velocity of a Kinematic Chain

In robotics, we often deal with a sequence of coordinate frames connected rigidly to each other, and we are interested in representing the angular velocities relative to a reference frame. Assume that two coordinate frames $o_1x_1y_1z_1$ and $o_2x_2y_2z_2$ are expressed to a reference frame $o_0x_0y_0z_0$ by R_1^0 and R_2^0, respectively. In case of all rotation matrices varying with time, we can say that

$$R_2^0(t) = R_1^0(t)R_2^1(t). \tag{5.24}$$

Taking the derivative of both sides of Eq. (5.24),

$$\dot{R}_2^0 = \dot{R}_1^0 R_2^1 + R_1^0 \dot{R}_2^1, \tag{5.25}$$

$$\dot{R}_2^0 = S(w_2^0) R_2^0, \tag{5.26}$$

$$\dot{R}_1^0 R_2^1 = S(w_1^0) R_1^0 R_2^1 = S(w_1^0) R_2^0, \tag{5.27}$$

$$R_1^0 \dot{R}_2^1 = R_1^0 S(w_2^1) R_2^1, \tag{5.28}$$

we can add identity matrix $I = R_1^{0T} R_1^0$ to Eq. (5.28), which does not affect its value

$$R_1^0 \dot{R}_2^1 = \underbrace{R_1^0 S(w_2^1) R_1^{0T}}_{S(R_1^0 w_2^1)} R_1^0 R_2^1, \tag{5.29}$$

$$R_1^0 \dot{R}_2^1 = S(R_1^0 w_2^1) R_2^0. \tag{5.30}$$

By substituting Eqs. (5.26), (5.27), and (5.30) into Eq. (5.25), we get

$$S(w_2^0) R_2^0 = S(w_1^0) R_2^0 + S(R_1^0 w_2^1) R_2^0. \tag{5.31}$$

Due to the linearity of the skew-symmetric matrix, Eq. (5.31) yields

$$w_2^0 = w_1^0 + R_1^0 w_2^1. \tag{5.32}$$

In general, we can say that

$$w_n^0 = w_1^0 + R_1^0 w_2^1 + R_2^0 w_3^2 + \cdots + R_{n-1}^0 w_n^{n-1}. \tag{5.33}$$

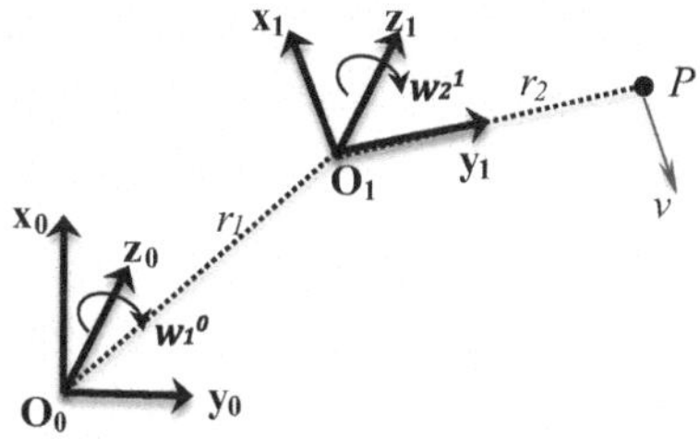

Fig. 5.4 Point on a kinematic chain subjected to angular and linear velocities

5.6 Linear Velocity

Equation (5.2) says that the linear velocity of a point rigidly attached to a body is just the cross product of the angular velocity of the body multiplied by the perpendicular distance from that point to the centre of rotation. The direction of the linear velocity will be in the tangential direction of the rotation. In this section, we will formulate a general expression for linear velocity for a point on a kinematic chain. To find the linear velocity of point P in Fig. 5.4, we will use the expression of Eq. (5.2).

$$v = w_2^1 \times r_2 + \dot{O}_1^0, \tag{5.34}$$

where $r_2 = R_1^0 P_2^1$ is the perpendicular distance from point P to O_1 expressed in the frame $o_0x_0y_0z_0$. The origin of frame 1 also has linear velocity expressed by the following equation:

$$\dot{O}_1^0 = w_1^0 \times r_1, \tag{5.35}$$

where r_1 is the position of O_1 expressed in the base frame or in robotics is just link length.

5.7 The Jacobian

Manipulator Jacobian or just Jacobian is a unique property for a specific robot manipulator. Jacobian is used to relate the velocities of the end-effector to the joint velocities. The dimensions of the Jacobian matrix are $6 \times n$, where n is the number of the links in the manipulator. Deriving the Jacobian matrix of a robot is very important, especially when we would like to find the dynamic equations of the robot. In this section, we will explain how to find the Jacobian considering the velocities of the end-effector, but we can consider calculation this matrix at any specific point on the robot. The Jacobian which is described in this section is joint variable dependent and is sometimes called the geometric Jacobian.

$$v_n^0 = J_v \dot{q}, \tag{5.36}$$

$$w_n^0 = J_w \dot{q}, \tag{5.37}$$

where $v = \begin{bmatrix} v_x & v_y & v_z \end{bmatrix}^T$ and $w = \begin{bmatrix} w_x & w_y & w_z \end{bmatrix}^T$.

$$H_n^0 = \begin{bmatrix} R_n^0 & O_n^0 \\ 0 & 1 \end{bmatrix}, \tag{5.38}$$

$$O_n^0 = \begin{bmatrix} x \\ y \\ z \end{bmatrix}, \tag{5.39}$$

$$J_v = \begin{bmatrix} \frac{\partial x}{\partial q_1} & \frac{\partial x}{\partial q_2} & \cdots & \frac{\partial x}{\partial q_n} \\ \frac{\partial y}{\partial q_1} & \frac{\partial y}{\partial q_2} & \cdots & \frac{\partial y}{\partial q_n} \\ \frac{\partial z}{\partial q_1} & \frac{\partial z}{\partial q_2} & \cdots & \frac{\partial z}{\partial q_n} \end{bmatrix}, \tag{5.40}$$

$$J_w = \begin{bmatrix} z_0^0 & z_1^0 & \cdots & z_{n-1}^0 \end{bmatrix}. \tag{5.41}$$

Equations (5.36) and (5.37) can be written in more concise manner as follows:

$$\xi = J\dot{q}, \tag{5.42}$$

where $\xi = \begin{bmatrix} v_n^0 & w_n^0 \end{bmatrix}^T$ is the body velocity and $J = \begin{bmatrix} J_v & J_w \end{bmatrix}^T$ is the Jacobian of the manipulator. Recall Eq. (3.8) and Eq. (3.27), where we have seen that the position vector is the first three elements of the fourth column of the transformation matrix while the z-vector is the first three elements of the third column of the homogenous transformation matrix.

For a prismatic joint, there is only translational movement, thus, there is no contribution to the orientation of the robot. Consequently, the z-vector of the prismatic joint is taken to be zero $\begin{bmatrix} 0 & 0 & 0 \end{bmatrix}^T$, as we will see in Example 5.1.

Example 5.1 SCARA Manipulator The SCARA manipulator has four degrees of freedom with a combination of three revolute joints and one prismatic joint. Figure 5.5 illustrates the frame assignment of the robot, while the spatial parameters are given in Table 5.1. By substitution of the spatial parameters in Eqs. (4.4) and (4.4a), we can find the homogenous transformation matrix for each link in the manipulator.

Equations (5.43)–(5.46) represent the four transformation matrices of the four links. We have to represent the orientation and position of each link to the reference frame; this what is happened in Eqs. (5.47) and (5.48), where we have calculated H_2^0, H_3^0, and H_4^0, respectively.

$$H_1^0 = \begin{bmatrix} C_1 & -S_1 & 0 & a_1C_1 \\ S_1 & C_1 & 0 & a_1S_1 \\ 0 & 0 & 1 & 0 \\ 0 & 0 & 0 & 1 \end{bmatrix}, \tag{5.43}$$

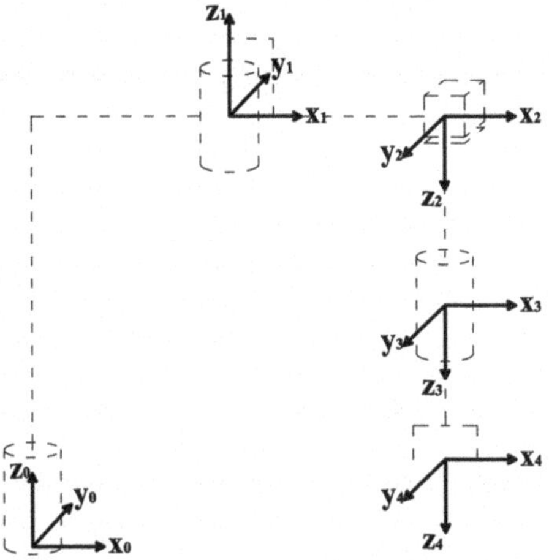

Fig. 5.5 SCARA manipulator

Table 5.1 Spatial parameters of the SCARA manipulator

Link ID	Frame	α	a	d	θ
1	$o_0x_0y_0z_0$–$o_1x_1y_1z_1$	0	a_1	0	θ_1
2	$o_1x_1y_1z_1$–$o_2x_2y_2z_2$	180	a_2	0	θ_2
3	$o_2x_2y_2z_2$–$o_3x_3y_3z_3$	0	0	P_3	—
4	$o_3x_3y_3z_3$–$o_4x_4y_4z_4$	0	0	d_4	θ_4

$$H_2^1 = \begin{bmatrix} C_2 & S_2 & 0 & a_2C_2 \\ S_2 & -C_2 & 0 & a_2S_2 \\ 0 & 0 & -1 & 0 \\ 0 & 0 & 0 & 1 \end{bmatrix}, \tag{5.44}$$

$$H_3^2 = \begin{bmatrix} 1 & 0 & 0 & 0 \\ 0 & 1 & 0 & 0 \\ 0 & 0 & 1 & P_3 \\ 0 & 0 & 0 & 1 \end{bmatrix}, \tag{5.45}$$

$$H_4^3 = \begin{bmatrix} C_1 & -S_1 & 0 & 0 \\ S_1 & C_1 & 0 & 0 \\ 0 & 0 & 1 & d_4 \\ 0 & 0 & 0 & 1 \end{bmatrix}, \tag{5.46}$$

$$H_2^0 = H_1^0 H_2^1 = \begin{bmatrix} C_{12} & S_{12} & 0 & a_1C_1 + a_2C_{12} \\ S_{12} & -C_{12} & 0 & a_1S_1 + a_2S_{12} \\ 0 & 0 & -1 & 0 \\ 0 & 0 & 0 & 1 \end{bmatrix}, \tag{5.47}$$

$$H_3^0 = H_1^0 H_2^1 H_3^2 = \begin{bmatrix} C_{12} & S_{12} & 0 & a_1C_1 + a_2C_{12} \\ S_{12} & -C_{12} & 0 & a_1S_1 + a_2S_{12} \\ 0 & 0 & -1 & P_3 \\ 0 & 0 & 0 & 1 \end{bmatrix}, \tag{5.48}$$

$$\begin{aligned} H_4^0 &= H_1^0 H_2^1 H_3^2 H_4^3 \\ &= \begin{bmatrix} C_4C_{12} + S_4S_{12} & C_4S_{12} - S_4C_{12} & 0 & a_1C_1 + a_2C_{12} \\ C_4S_{12} - S_4C_{12} & -C_4C_{12} - S_4S_{12} & 0 & a_1S_1 + a_2S_{12} \\ 0 & 0 & -1 & -d_3 - P_3 \\ 0 & 0 & 0 & 1 \end{bmatrix} \end{aligned} \tag{5.49}$$

Now, from Eq. (5.49) we can find the translational part of the Jacobian J_v as follows:

$$J_v = \begin{bmatrix} \frac{\partial x}{\partial \theta_1} & \frac{\partial x}{\partial \theta_2} & \frac{\partial x}{\partial P_3} & \frac{\partial x}{\partial \theta_4} \\ \frac{\partial y}{\partial \theta_1} & \frac{\partial y}{\partial \theta_2} & \frac{\partial y}{\partial P_3} & \frac{\partial y}{\partial \theta_4} \\ \frac{\partial z}{\partial \theta_1} & \frac{\partial z}{\partial \theta_2} & \frac{\partial z}{\partial P_3} & \frac{\partial z}{\partial \theta_4} \end{bmatrix}, \tag{5.50}$$

where x, y, and z are the first three elements in the fourth column of Eq. (5.49). By taking the partial derivative, we have the following:

$$J_v = \begin{bmatrix} -a_1S_1 - a_2S_{12} & -a_2S_{12} & 0 & 0 \\ a_1C_1 + a_2C_{12} & a_2C_{12} & 0 & 0 \\ 0 & 0 & -1 & 0 \end{bmatrix}. \tag{5.51}$$

The rotational part of the Jacobian J_w is the z-vector of all of the coordinate frames represented to the reference frame

$$J_w = \begin{bmatrix} z_0^0 & z_1^0 & \underbrace{z_2^0}_{\text{prismatic joint}} & z_3^0 \end{bmatrix} = \begin{bmatrix} z_0^0 & z_1^0 & 0 & z_3^0 \end{bmatrix}, \tag{5.52}$$

where $z_0^0 = \begin{bmatrix} 0 & 0 & 1 \end{bmatrix}^T$ while z_1^0 and z_3^0 are the first three elements of the third columns of Eqs. (5.43) and (5.48), respectively. Substituting all the vectors yields

$$J_w = \begin{bmatrix} 0 & 0 & 0 & 0 \\ 0 & 0 & 0 & 0 \\ 1 & 1 & 0 & -1 \end{bmatrix}. \tag{5.53}$$

Combining the translational part and rotational part we can find the overall Jacobian matrix

$$J = \begin{bmatrix} -a_1S_1 - a_2S_{12} & -a_2S_{12} & 0 & 0 \\ a_1C_1 + a_2C_{12} & a_2C_{12} & 0 & 0 \\ 0 & 0 & -1 & 0 \\ 0 & 0 & 0 & 0 \\ 0 & 0 & 0 & 0 \\ 1 & 1 & 0 & -1 \end{bmatrix}. \tag{5.54}$$

Example 5.2 Figure 5.6 shows a model of a 5DOF revolute-joint manipulator, with its spatial parameters shown in Table 5.2. We will use Symbolic Math Toolbox in MATLAB to find the Jacobian of the manipulator. This is a generic solution that can be modified to find the Jacobian of any manipulator and will be extended later in Chap. 7 to find the dynamic equations using a programming procedure. The usual work, which is the same as the previous example, is used to build the code, and the details are left for the reader to understand the code.

Code:

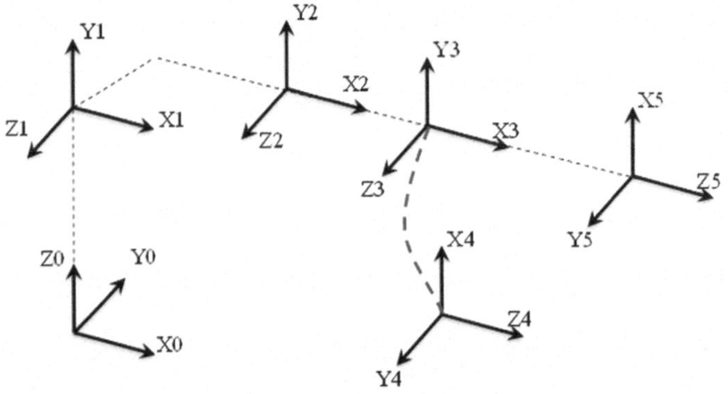

Fig. 5.6 Five-degree-of-freedom 5R robot manipulator

Table 5.2 Spatial parameters of the 5R manipulator

Link ID	Frame	α	a	d	θ
1	$o_0x_0y_0z_0$–$o_1x_1y_1z_1$	90	0	d_1	θ_1
2	$o_1x_1y_1z_1$–$o_2x_2y_2z_2$	0	a_2	d_2	θ_2
3	$o_2x_2y_2z_2$–$o_3x_3y_3z_3$	0	a_3	0	θ_3
4	$o_3x_3y_3z_3$–$o_4x_4y_4z_4$	90	0	0	θ_4
5	$o_4x_4y_4z_4$–$o_5x_5y_5z_5$	0	0	d_5	θ_5

```
clear;
clc;
format short G
syms t1(t) t2(t) t3(t) t4(t) t5(t) d1 a2 a3 d5

th1= t1;                % joint angle 1
tw_1 = 90;              % twist angle for Link 0
a_1 = 0;                % distance between z0 and z1
d_1 = d1;               % distance between x0 and x1
th2= t2;                % joint angle 2
tw_2 = 0;               % twist angle for Link 1
a_2 = a2;               % distance between z1 and z2
d_2 = d2;               % distance between x1 and x2
th3= t3;                % joint angle 3
tw_3 = 0;               % twist angle for Link 2
a_3 = a3;               % distance between z2 and z3
d_3 = 0;                % distance between x2 and x3
th4= t4;                % joint angle 4
tw_4 = 90;              % twist angle for Link 3
a_4 = 0;                % distance between z3 and z3
d_4 = 0;                % distance between x3 and x4
th5= t5;                % joint angle 5
tw_5 = 0;               % twist angle for Link 4
a_5 = 0;                % distance between z4 and z5
d_5 = d5;               % distance between x4 and x5

H_1= Denavit(th1,tw_1,a_1,d_1);
H_2= Denavit(th2,tw_2,a_2,d_2);
H_3= Denavit(th3,tw_3,a_3,d_3);
H_4= Denavit(th4,tw_4,a_4,d_4);
H_5= Denavit(th5,tw_5,a_5,d_5);
%%%%%%%%% transformation matrices at center of masses
CN1 = H_1;
CN2=H_1*H_2;
CN3=H_1*H_2*H_3;
CN4=H_1*H_2*H_3*H_4;
CN5=H_1*H_2*H_3*H_4*H_5;
%%%%%%%%%%%%%%%%%%%%%
cc1 = CN1(t); % clone to enable indexing of symbolic
```

```
function
cc2 = CN2(t); % clone to enable indexing of symbolic
function
cc3 = CN3(t); % clone to enable indexing of symbolic
function
cc4 = CN4(t); % clone to enable indexing of symbolic
function
cc5 = CN5(t); % clone to enable indexing of symbolic
function
%%%%%%%%%%%%%%%%%%%%%%%%%%%%%%%%%%%%%%%%%%%%%%%%%%%%%%%%
Jw1 =[0;0;1];
Jw2 =cc1(1:3,3);
Jw3 =cc2(1:3,3);
Jw4 =cc3(1:3,3);
Jw5 =cc4(1:3,3);
Jw=[Jw1,Jw2,Jw3,Jw4,Jw5];
%%%%%%%%%%%%%%%%%%%%%%%%%%%%%%%%%%%%%%%%%%%%%%%%%%%%%%%%
Jv =
[functionalDerivative(cc5(1,4),th1),functionalDerivativ
e(cc5(1,4),th2),...

functionalDerivative(cc5(1,4),th3),functionalDerivative
(cc5(1,4),th4),...
    functionalDerivative(cc5(1,4),th5);...

functionalDerivative(cc5(2,4),th1),functionalDerivative
(cc5(2,4),th2),...

functionalDerivative(cc5(2,4),th3),functionalDerivative
(cc5(2,4),th4),...
    functionalDerivative(cc5(2,4),th5);...

functionalDerivative(cc5(3,4),th1),functionalDerivative
(cc5(3,4),th2),...

functionalDerivative(cc5(3,4),th3),functionalDerivative
(cc5(3,4),th4),...
    functionalDerivative(cc5(3,4),th5)];
J = [Jv;Jw]; % Jacobian matrix
```

The following code is for Denavit function

```
function [DH] = Denavit(theta,alpha,a,d)
DH = [cos(theta),-
1*sin(theta)*cosd(alpha),sin(theta)*sind(alpha),a*cos(t
heta);...
    sin(theta),cos(theta)*cosd(alpha),-
1*cos(theta)*sind(alpha),a*sin(theta);...
    0,sind(alpha),cosd(alpha),d;...
    0,0,0,1];
end
```

5.8 Inverse Acceleration and Velocity

We have seen that Eq. (5.42) provides a mapping between the end-effector velocity vector and joint velocities. Given the joint velocity vector, we can find the corresponding end-effector velocity vector. The inverse problem, when the end-effector velocity is given and the corresponding joint velocities are required, is quite an easy problem for $n \times n$ Jacobian:

$$\dot{q} = J^{-1}\xi, \tag{5.55}$$

Equation (5.55) is valid only for a square Jacobian matrix because this is the only case in which we can invert the matrix. For nonsquare Jacobian, the inverse problem could be a problem and there are some algebraic solutions for some cases, but we will not deal with such solutions in this text. In this section, we will derive a general fitness function for square and nonsquare Jacobian matrices that can be minimized to find the inverse velocity and acceleration. For a given velocity vector of the end-effector ξ_g and Jacobian J, consider:

$$\dot{q}_O = \left[\dot{q}_1\ \dot{q}_2 \cdots \dot{q}_n\right], \tag{5.56}$$

where O denotes the optimized joint velocity vector,

such that

$$\xi_O = J\dot{q}_O, \tag{5.57}$$

where ξ_O represents the optimized body velocity vector during the run-time of the optimization algorithm. The minimum difference between the given body velocity and the optimized body velocity is the fitness function that we should consider to minimize

$$\text{fitness} = \left\| \xi_g - \xi_O \right\|. \tag{5.58}$$

By the same logic, we can speak about the inverse acceleration of the manipulator where end-effector accelerations are given and the required information is the corresponding joint acceleration vector. The time derivative of both sides of Eq. (5.42) yields

$$\ddot{X} = \dot{J}\dot{q} + J\ddot{q}, \tag{5.59}$$

where $\ddot{X}$ is the body acceleration vector.

$$\ddot{X} = \left[a_x \; a_y \; a_z \; \varpi_x \; \varpi_y \; \varpi_z \right]^T, \tag{5.60}$$

where a and ϖ are the linear and angular velocity of the end-effector, respectively. $\dot{J}$ is the time derivative of the Jacobian considering that $J = J(q_i(t))$. To find the inverse acceleration, we may follow the same procedure.

Consider:

$$\ddot{q}_O = \left[\ddot{q}_1 \; \ddot{q}_2 \cdots \ddot{q}_n \right],$$

$$\dot{q}_O = \left[\dot{q}_1 \; \dot{q}_2 \cdots \dot{q}_n \right], \tag{5.61}$$

such that

$$\ddot{X}_O = \dot{J}\dot{q}_O + J\ddot{q}_O(5.54). \tag{5.62}$$

Thus, the function that should be minimized is

$$\text{fitness} = \left\| \ddot{X}_g - \ddot{X}_O \right\|. \tag{5.63}$$

where $\ddot{X}_g$ is the given body acceleration vector.

Example 5.3 Derive the body acceleration vector of the manipulator in Example 5.2.

We have modified the code solution of example 5.2 to be appropriate for finding the forward acceleration calculations. Note that the MATLAB function *diff()* returns the time derivative of the variable which is used to find the time derivative of the Jacobian.

```
clear;
clc;
format short G
syms t1(t) t2(t) t3(t) t4(t) t5(t) d1 d2 a2 a3 d5
syms qd1 qd2 qd3 qd4 qd5            % Joints angular
velocity vector
syms q2d1 q2d2 q2d3 q2d4 q2d5    % Joints angular
acceleration vector
qd  = [qd1;qd2;qd3;qd4;qd5];
q2d = [q2d1;q2d2;q2d3;q2d4;q2d5];

th1= t1;                   % joint angle 1
tw_1 = 90;                 % twist angle for Link 0
a_1 = 0;                   % distance between z0 and z1
d_1 = d1;                  % distance between x0 and x1
th2= t2;                   % joint angle 2
tw_2 = 0;                  % twist angle for Link 1
a_2 = a2;                  % distance between z1 and z2
d_2 = d2;              % distance between x1 and x2
th3= t3;                   % joint angle 3
tw_3 = 0;                  % twist angle for Link 2
a_3 = a3;                  % distance between z2 and z3
d_3 = 0;                   % distance between x2 and x3
th4= t4;                   % joint angle 4
tw_4 = 90;                 % twist angle for Link 3
a_4 = 0;                   % distance between z3 and z3
d_4 = 0;                   % distance between x3 and x4
th5= t5;                   % joint angle 5
tw_5 = 0;                  % twist angle for Link 4
a_5 = 0;                   % distance between z4 and z5
d_5 = d5;                  % distance between x4 and x5
H_1= Denavit(th1,tw_1,a_1,d_1);
H_2= Denavit(th2,tw_2,a_2,d_2);
H_3= Denavit(th3,tw_3,a_3,d_3);
H_4= Denavit(th4,tw_4,a_4,d_4);
H_5= Denavit(th5,tw_5,a_5,d_5);
%%%%%%%%%%%%%%%%%%% transformation matrices at center
of masses
CN1 = H_1;
CN2=H_1*H_2;
CN3=H_1*H_2*H_3;
CN4=H_1*H_2*H_3*H_4;
CN5=H_1*H_2*H_3*H_4*H_5;
%%%%%%%%%%%%%%%%%%%%%%
cc1 = CN1(t); % clone to enable indexing of symbolic
```

```
function
cc2 = CN2(t); % clone to enable idexing of symbolic
function
cc3 = CN3(t); % clone to enable idexing of symbolic
function
cc4 = CN4(t); % clone to enable idexing of symbolic
function
cc5 = CN5(t); % clone to enable idexing of symbolic
function
%%%%%%%%%%%%%%%%%%%%%%%%%%%%%%%%%%%%%%%%%%%%%%%%%%%%%%%%%%%%%%%%
%%
Jw1 = [0;0;1];
Jw2 =cc1(1:3,3);
Jw3 =cc2(1:3,3);
Jw4 =cc3(1:3,3);
Jw5 =cc4(1:3,3);
Jw=[Jw1,Jw2,Jw3,Jw4,Jw5];
%%%%%%%%%%%%%%%%%%%%%%%%%%%%%%%%%%%%%%%%%%%%%%%%%%%%%%%%%%%%%%%%
%%
Jv =
[functionalDerivative(cc5(1,4),th1),functionalDerivativ
e(cc5(1,4),th2),...

functionalDerivative(cc5(1,4),th3),functionalDerivative
(cc5(1,4),th4),...
    functionalDerivative(cc5(1,4),th5);...

functionalDerivative(cc5(2,4),th1),functionalDerivative
(cc5(2,4),th2),...

functionalDerivative(cc5(2,4),th3),functionalDerivative
(cc5(2,4),th4),...
    functionalDerivative(cc5(2,4),th5);...

functionalDerivative(cc5(3,4),th1),functionalDerivative
(cc5(3,4),th2),...

functionalDerivative(cc5(3,4),th3),functionalDerivative
(cc5(3,4),th4),...
    functionalDerivative(cc5(3,4),th5)];
J = [Jv;Jw];
Jd = diff(J(t)); % clone
JJ = J(t);       % clone
X = Jd*qd+JJ*q2d
```

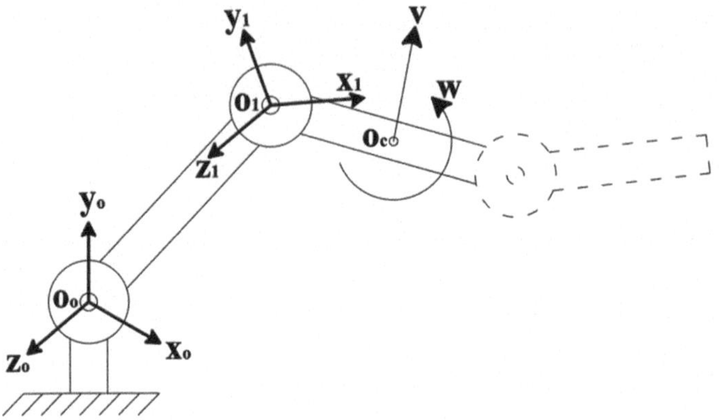

Fig. 5.7 RRR planar manipulator

5.9 Additive to the Jacobian

As we have explained before, the Jacobian provides a mapping for the velocities of a point on the manipulator to the joint space. In the previous examples, we considered the end-effector velocity and acceleration. In fact, it is more important to describe the velocity and acceleration of different points on the manipulator rather than on its end-effector, especially when dynamic analysis is required. Consider Fig. 5.7, which shows a 3DOF revolute-joint manipulator. The centre of mass of the link 2 is denoted by o_c with linear velocity v and angular velocity w. We can derive the Jacobian of the manipulator at the centre of mass of link 2 directly to represent its velocity and acceleration to the actuators at joint 0 and joint 1. Note that link 3 does not affect the velocity of the link 2 kinematically, but the reaction from there can affect the link when we consider kinetic analysis, as we will explain in Chap. 7.

5.10 Summary

In this chapter, we have learned how to find the relationship among velocity components of the end-effector and velocities of the joints variables. This relationship is the Jacobian matrix which is driven from forward kinematic equations. Jacobian is very important when we would like to describe the velocities of the centre of masses of the links and end-effector in order to formulate the dynamic equations as we will in Chap. 7.

Exercises

5.1 Derive the Jacobian of the two-link planar manipulator in Example 4.1.

5.2 For the same manipulator in Problem 5.1, derive the velocity and acceleration kinematics of the robot.

5.3 For a given velocity vector of the end-effector of the LabVolt 5150 manipulator, derive the fitness function for the inverse velocity. It can be optimized by one of the optimization algorithms of your choice.

5.4 Relate the velocity and acceleration of the centre of mass of link 2 of the manipulator shown in Fig. 5.7 to the joint space.

Chapter 6
Path and Trajectory Planning

6.1 Introduction

In the previous chapters, we have studied the geometry of the robot and the kinematic solutions associated with that geometry without giving any attention to the constraints imposed by the workspace. We have developed solutions for the forward kinematics and inverse kinematics of the robot manipulator, assuming that we have free space surrounding the robot and its possible configurations. For a workspace with constraints like other robots, equipment, workers, etc., it is not feasible to find a specific configuration without considering the environment that the robot works in Fig. 6.1 illustrates a space already occupied by several other objects to be avoided. In this chapter, we will study collision-free path planning to describe the geometric motion of the whole robot or one of its parts.

Path planning is one of the most challenging problems in computer science, where many algorithms have been proposed to solve it. We will describe the most popular algorithms for path planning with a detailed description of their coding. Path planning describes the motion geometrically, while trajectory planning describes the velocity, acceleration, and forces on that path. This chapter also presents the issue of trajectory planning with an example of applied software.

6.2 Path Planning

Path planning is the problem of finding the shortest path between two given points where the whole robot or one of its parts can follow this path. Axiomatically, the shortest path between two points in Cartesian space is the straight line connecting these points. However, the problem gets complicated while working in a constrained environment with static or dynamic obstacles. If the configuration space is denoted by Q, then we can say that the taboo configuration Qt is a set of configurations that bring the robot into contact with obstacles O_i and the possible configurations Qp are

H. N. Ghafil and K. Jármai (ed.), *Optimization for Robot Modelling with MATLAB*,
https://doi.org/10.1007/978-3-030-40410-9_6

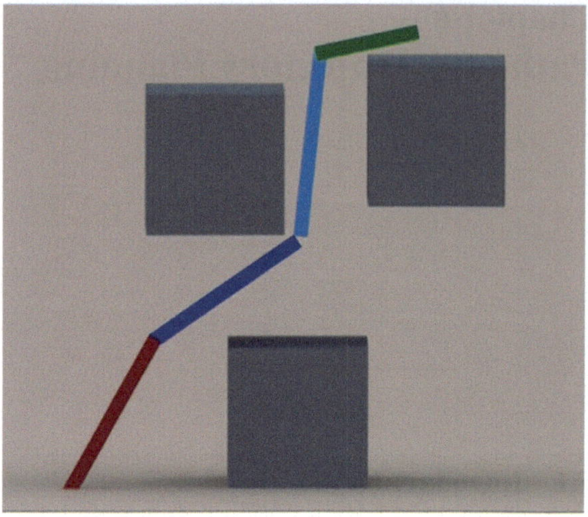

Fig. 6.1 Manipulator while avoiding obstacles

$$Qp = Q - Qt. \tag{6.1}$$

In other words, we can say that

$$Qt = \{q_i \in Q | Q(q_i) \cap O_i \neq \phi\}$$

According to the previous definitions, motion planning can be defined as the set of all possible configurations that do not collide with obstacles, while the whole robot or some parts of the robot follow a specific path. Planning a collision-free path is a geometric issue where this path is merely a set of ordered Cartesian points in the space. There is no description or information about how the robot follows the path point by point. Trajectory planning describes the motion of the robot on a collision-free path and gives information about the instantaneous position, velocity, and acceleration. In other words, trajectory planning is the next operation that should be done after path planning. In the following sections, we will describe some of the well-known techniques to plan the optimal path in a constrained workspace.

6.3 Ant Colony Optimization for TSP

The travelling salesman problem (TSP) is nondeterministic polynomial time problem that has been widely studied and has become a benchmark for the efficiency of the optimization algorithms. TSP was formulated in 1930 and asks the question, 'Given the distances among each pair of points in space, what is the shortest path that visits all the points and goes back to the origin'? Many applications have been proposed for this question, like considering the points as cities to be visited or customers in restaurant to be served. The solution for this issue is properly ordered points selected from the given points. This kind of problem is called discrete optimization. In Sect. 2.2.1, we described ant colony optimization (ACO), which is suitable for discrete optimization and should be perfect for solving TSP.

TSP is a discrete path planning problem where we have to find the shortest route among many others that visit all the cities or points. To be consistent with the definition of path planning, we will modify the question of the travelling salesman problem a bit. Starting from a specific start point, what is the shortest path that visits all the points with given distances among them? The optimal solution for this question imposes that the last point on the path should be the goal point. Later, we shall give hints to the reader how to solve the classical problem.

Consider the following data represented in Fig. 6.2. It is required to find the shortest path that starts from the point designated by the circle and travels through all other points. There are many possible routes for this problem, and we will find the best one by using ACO with the following code example.

Example 6.1 We are given the following Cartesian coordinates for a set of points:

(20,60), (90,70), (15,90), (83,100), (60,65), (10,70), (30,80), (60,40), (100,70), (90,20), (95,30), (50,5), (90,13), (40,66), (85,37), (87,90), (80,7). The solution required is the optimum path that starts from (20,60) and travels through all other points.

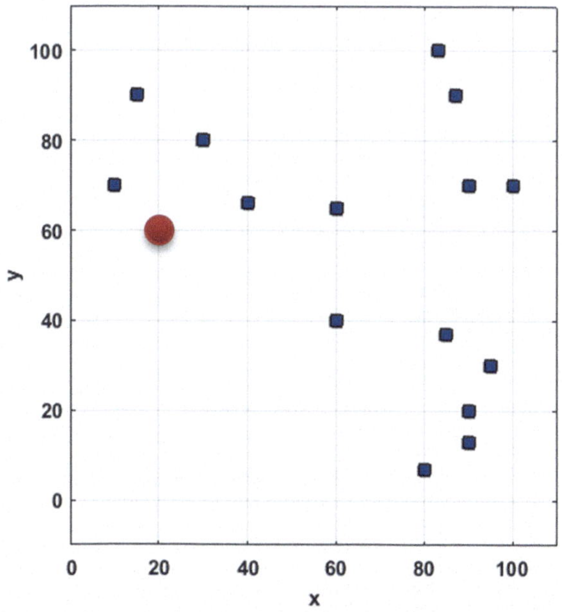

Fig. 6.2 Example of randomly distributed points that have to be visited by one best route

We tend to split the code of the solution into separated functions (m-files) for easier understanding. Firstly, we start by creating the space of the problem that holds all the information about the given point. We shall call it `problemSpace()`:

```
function space=problemSpace()
    x=[20 90 15 83  60  10  30 60 100 90  95 50 90   40
85 87 80];% x-axis of the points
    y=[60 70 90 100 65  70  80 40 70  20  30 5  13   66
37 90 7]; % y-axis of the points
    for i=1:length(x)-1
        for j=i+1:length(x)
            H(i,j)=sqrt((x(i)-x(j))^2+(y(i)-y(j))^2);
            H(j,i)=D(i,j);
        end
    end
    space.x=x;
    space.y=y;
    space.H=H;
end
```

Notice that matrix H is the heuristic information that holds the distance between every two points in the set, see Sect. 2.2.1. For instance, distance from point 1 to point 2 is the same as the distance between point 2 and point 1, and that is the reason why $H(i,j) = H(j,i)$.

The following code is for the main ACO algorithm:

```
clc;
clear;
close all;
%% Problem Definition
space= problemSpace();
CostFunction=@(path) cost(path,space);
Npoint = length(space.x);
%% ACO Parameters
NIter = 250;          % Maximum Number of Iterations
NAnt = 50;            % Number of Ants
tu0 = 0.01;       % Initial value for the Pheromone
alpha =1;           % Pheromone  Weight
beta =1;            % Heuristic Weight
ro = 0.07;          % Rate of Evaporation
%% Initialization
```

```
ita=1./space.H;              % Heuristic Information
Matrix
tu=tu0*ones(Npoint,Npoint);    % Pheromone Matrix
costs=zeros(NIter,1);     % vector for all Best Cost
Values
% Empty Ant
templet.Path=[];
templet.Cost=[];
% Ant Colony Matrix
ant = repmat(templet,NAnt,1);
% Best Ant
Bestpath.Cost=inf;
%% ACO Main Loop
for iter=1:NIter
    % Move Ants
    for k=1:NAnt
```

In the following line, we have forced all ants to start their paths from point 1, then to select the next point on the path according to Eq. (2.1).

```
        ant(k).Path = 1;
        for l=2:Npoint
            i=ant(k).Path(end);
            P=tu(i,:).^alpha.*ita(i,:).^beta;
            P(ant(k).Path)=0;
            P=P/sum(P);     % equation (2.1)
            j=RouletteWheelSelection(P);
            ant(k).Path=[ant(k).Path j];
        end
        ant(k).Cost=CostFunction(ant(k).Path);
        if ant(k).Cost<Bestpath.Cost
            Bestpath=ant(k);
        end
    end
    % Update Pheromones
    for k=1:NAnt
        path=ant(k).Path;
        for l=1:Npoint-1
            i=path(l);
            j=path(l+1);
            tu(i,j)=tu(i,j)+ 1/ant(k).Cost;
%equation (2.3)
        end
```

```
    end
    % Evaporation
    tu=(1-ro)*tu;  % part of equation (2.3)
    % Store Best Cost
    costs(iter)=Bestpath.Cost;
    % Show Iteration Information
    disp(['Iteration = ' num2str(iter) ': Cost = '
num2str(costs(iter))]);
    % Plot Solution
    figure(1);
    plotpath(Bestpath.Path,space);
    pause(0.01);
end
```

The following code is for the cost function, which is the length of the instantaneous path proposed by an ant.

```
function L=cost(path,space)

    n=length(path);
    L=0;
    for i=1:n-1
        L=L+space.H(path(i),path(i+1));
    end

end
```

Roulette wheel selection function, which is described in Sect. 2.21.2.4, is represented by the following.

```
function j=RouletteWheelSelection(P)
    r=rand;
    C=cumsum(P);
    j=find(r<=C,1);
end
```

To plot the best path for each iteration, we have used the following code snippets:

```
function plotpath(path,space)
    plot(space.x(path),space.y(path),'k-s',...
        'MarkerSize',9,...
        'MarkerFaceColor','b',...
        'LineWidth',1.3);
    xlabel('x');
    ylabel('y');
    axis ([0,110,0,110]);
    grid on;
end
```

Many works have dealt with path planning in a technique similar to the A-star algorithm, which depends on dividing the workspace into a grid and the ants have to walk on that grid from a given start point to goal point. For constrained path planning, grid elements that belong to the obstacles should have an infinite value of heuristic information to ensure that ants will never choose it for their path to the goal. For details, see Brand et al. (2010) or Raja and Pugazhenthi (2012).

6.4 Cubic Spline Curves

Spline function (Cheney and Kincaid 2012) is a polynomial function that was used originally for approximation problems. The pieces of this polynomial provide a smooth connection between two points via some control point. In other words, control points are responsible for the value and shape of the spline function that affects the interpolated values. Figure 6.3, as an example, shows how the spline function changes when changing the coordinates of the control points (which are expressed as circles), while ends of the spline are fixed in all cases at $S1$ and $S2$. We can say that at fixed terminal ends, control points are only the affecting parameters on the final topology of the graph.

From the previous brief explanation about the spline curve and its control points, you can see that it is exactly what we need for our task on path planning. We have a fixed start point, fixed target point, and a collision-free path between them, and this coincides with the behaviour of the spline function. For more details, see Fig. 6.4a. We can conclude that there is an infinite number of splines (paths) connecting the same terminal ends. The proposed path is formed by specific Cartesian coordinates of the control points, and there should be only one optimal path which is the shortest distance between the ends of the graph. The infinite number of possible paths comes from the continuity of the position variable of the control points.

Thus, the optimum path is formed by optimum selection of the coordinates of the control points. Assuming that we have three control points, then we can express the optimization vector for this problem by six variables in a two-dimensional workspace and nine variables in the case of a three-dimensional workspace. Notice that each Cartesian point is represented by $\begin{bmatrix} x & y \end{bmatrix}^T$ and $\begin{bmatrix} x & y & z \end{bmatrix}^T$ in 2D and 3D, respectively.

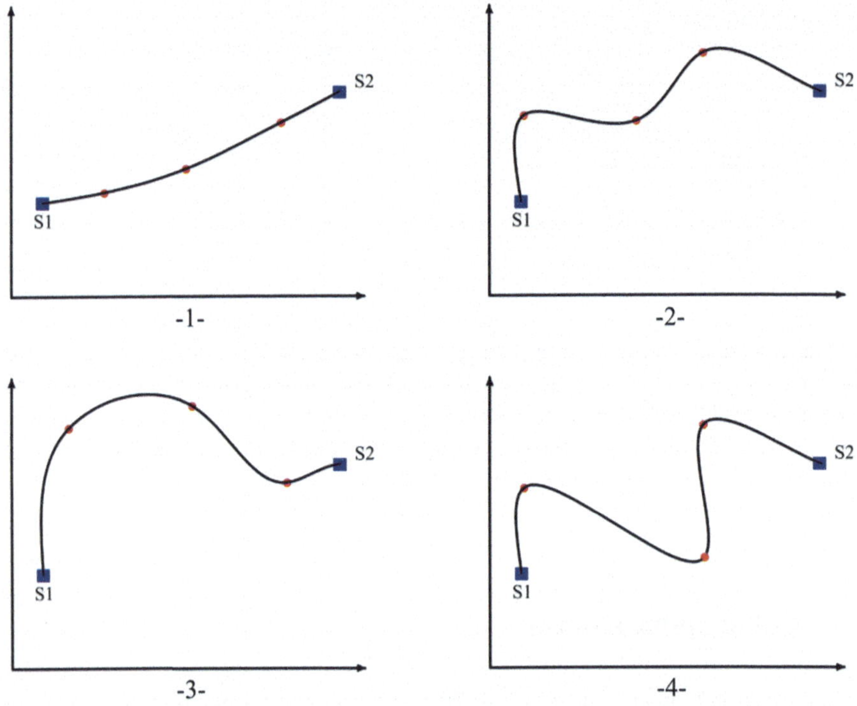

Fig. 6.3 Four different samples of spline function due to changing the position of the control points (circles), while ends of the spline are fixed at $S1$ and $S2$

The constraints in the working area should be imposed on the optimized vector, which path should never come into contact with any obstacle (unlike in Fig. 6.4b). The simple way is to apply a static penalty on the solution vector that forms a graph which touches an obstacle. This penalty is enough to get rid of that solution and force the optimization algorithm to propose another feasible one. We will show how to detect a collision path and apply the penalty in the following example.

6.5 Collision Detection

In this section, collision detection between path and obstacles will be described in detail. Assume that we have a path consisting of successive points p_i, $i = 1 \ldots n$ where n is the total number of points on the path as shown in Fig. 6.5. Also, there are O_j obstacles with radius $r_j, j = 1 \ldots k$ where k is the total number of obstacles in the workspace. A collision will occur when the distance between pi and the centre of the obstacle O_j is less than the radius r_j.

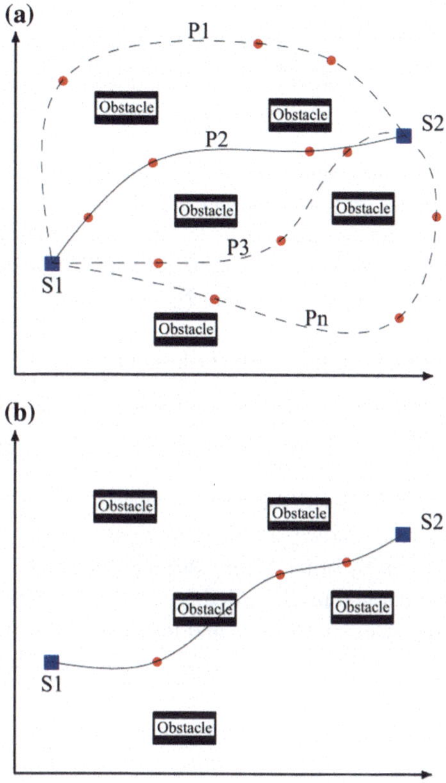

Fig. 6.4 **a** Infinite number of possible paths between two given points, **b** sample prohibited path contacting at least one obstacle

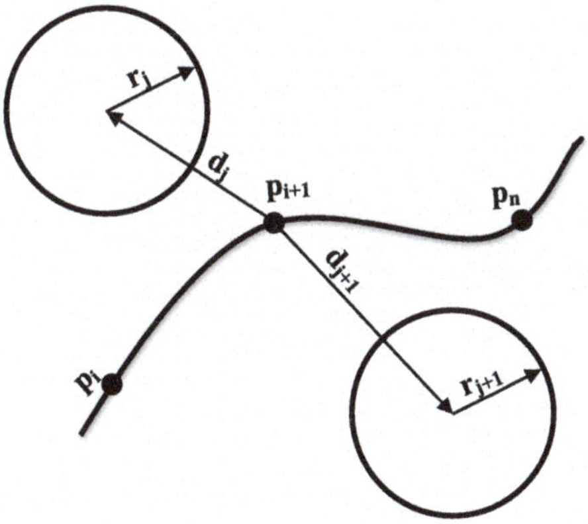

Fig. 6.5 Circular obstacles detection

$$d_j = \sqrt{(px_i - Ox_j)^2 + (py_i - Oy_j)^2}, \tag{6.2}$$

$$d_j < r_j, \tag{6.3}$$

where x and y denote the x-axis and y-axis of p_i and O_j, respectively.

For rectangular obstacles, any point p_i on the path should not be contained within the space occupied by the rectangular obstacle. The rectangular object is defined by a beginning point in a space B, width w, and height H. A collision occurs when the x-axis of p_i (px_i) and y-axis of p_i (py_i) within $B_x + w$ and $B_y + H$, respectively.

Example 6.2 Starting from $\begin{bmatrix} 0 & 0 & 0 \end{bmatrix}^T$ to the task point $\begin{bmatrix} 4 & 6 & 8 \end{bmatrix}^T$, it is required to find the shortest path in a constrained space with four spherical obstacles at $\begin{bmatrix} 1.5 & 4.5 & 2 \end{bmatrix}^T$, $\begin{bmatrix} 4 & 3 & 4 \end{bmatrix}^T$, $\begin{bmatrix} 1.2 & 1.5 & 2.5 \end{bmatrix}^T$, and $\begin{bmatrix} 2 & 4 & 6 \end{bmatrix}^T$ having radii 1.5, 1, 0.8, and 1.5, respectively.

For this problem, we have used the particle swarm operation (PSO) algorithm described in Sect. 2.21.1, and however, the reader can use the same procedure and function to find the solution by other optimization algorithms.

Firstly, we define a separated m file called `problemDefin()` to hold all of the information about workspace, constraints, and terminals of the path.

```
function problem = problemDefin()
    % start point
    xs=0;
    ys=0;
    zs=0;
    % Goal point
    xt=4;
    yt=6;
    zt=8;
    % Coordinates of the spherical obstacles
    xob=[1.5 4.0 1.2 2];
    yob=[4.5 3.0 1.5 4];
    zob=[2 4.0 2.5 6];
    % Radii of the spherical obstacles
    rob=[1.5 1.0 0.8 1.5];
```

```
    % Boundaries of the workspace
    xmin=-10; % x-direction
    xmax= 10;
    ymin=-10; % y-direction
    ymax= 10;
    zmin=-10; % z-direction
    zmax= 10;
    % Building structure variables that holds all the
information about
    % the problem
    problem.xs=xs;
    problem.ys=ys;
    problem.zs=zs;
    problem.xt=xt;
    problem.yt=yt;
    problem.zt=zt;
    problem.xob=xob;
    problem.yob=yob;
    problem.zob=zob;
    problem.rob=rob;
    problem.xmin=xmin;
    problem.xmax=xmax;
    problem.ymin=ymin;
    problem.ymax=ymax;
    problem.zmin=zmin;
    problem.zmax=zmax;
end
```

The cost function that should be optimized

```
function [cost, path]= Cost(c_points,problem)
    x=c_points.x;
    y=c_points.y;
    z=c_points.z;

    xs=problem.xs;
    ys=problem.ys;
    zs=problem.zs;
    xt=problem.xt;
    yt=problem.yt;
    zt=problem.zt;
    xob=problem.xob;
```

```
yob=problem.yob;
zob=problem.zob;
rob=problem.rob;

X=[xs x xt];
Y=[ys y yt];
Z=[zs z zt];

T=linspace(0,1,length(X));
tt=linspace(0,1,50);
xx=spline(T,X,tt);
yy=spline(T,Y,tt);
zz=spline(T,Z,tt);
```

xx, *yy*, and *zz* are the *x*, *y*, and *z* axes of the fifty points which are equally distributed along the cubic spline. Later, we will calculate the distances between each successive point, which represents the overall length of the path.

```
dx=diff(xx);
dy=diff(yy);
dz=diff(zz);
path_length=sum(sqrt(dx.^2+dy.^2+dz.^2));
```

As we have explained, an undesirable path is one that collides with at least one obstacle. Thus, we shall impose a penalty on the undesirable path by saying that if the distance between any point on the spline is less than the radius of any obstacle, then a penalty of a large integer number, for example, 100 or 1000, should be added to the length of the proposed path. If there is no contact with any obstacle, then the penalty will have zero value. The final value, which is a composite of path length plus penalty values, is the cost function that should be in the minimum value. This technique is called static penalty and is usually used for constrained optimization problems. In other words, the following code is a simple technique for collision detection which depends on adding a static penalty using sequential programming steps.

```
penalty = 0;
for k=1:length(rob)
    for m=1:length(xx)
    d=sqrt((xx(m)-xob(k)).^2+(yy(m)-
yob(k)).^2+(zz(m)-zob(k)).^2);
   if(d<rob(k))
      penalty=penalty+100;
   end

    end
end
```

```
    path.X=X;
    path.Y=Y;
    path.Z=Z;
    path.xx=xx;
    path.yy=yy;
    path.zz=zz;
    path.path_length=path_length;
    path.penalty=penalty;
    cost=path.path_length + path.penalty;

end
```

Main PSO algorithm

```
clc;
clear;
close all;
%%  Definition
problem = problemDefin();
costfun=@(x) Cost(x,problem);     % Cost Function
N_var = 3;          % Number of control points on the
curve
VarSize=[1 N_var];    % Size of Decision Variables
Matrix
% Boundaries of the working area
CoordMin.x=problem.xmin;              % Minimum value in
x-direction
CoordMax.x=problem.xmax;              % Maximum value in
x-direction
CoordMin.y=problem.ymin;              % Minimum value in
y-direction
CoordMax.y=problem.ymax;              % Maximum value in
y-direction
CoordMin.z=problem.zmin;              % Minimum value in
z-direction
CoordMax.z=problem.zmax;              % Maximum value in
z-direction
%% PSO Parameters
Iter  = 150;            % Number of Iterations
N_pop = 50;             % Population Size
w=1;                  % Inertia Weight
wdamp=0.95;           % Inertia Weight Damping Ratio
c1=1.5;               % Personal Learning Coefficient
c2=1.5;               % Global Learning Coefficient
%% Initialization
% Create Empty Particle Structure
control_point.Position=[];
```

```
control_point.Velocity=[];
control_point.Cost=[];
control_point.Sol=[];
control_point.Best.Position=[];
control_point.Best.Cost=[];
control_point.Best.Sol=[];
% Initialize Global Best
GlobalBest.Cost=inf;
% Create Particles Matrix
C_points=repmat(control_point,N_pop,1);
% Initialization Loop
for i=1:N_pop
    % Initialize Position
    C_points(i).Position.x =
unifrnd(CoordMin.x,CoordMax.x,VarSize);
    C_points(i).Position.y =
unifrnd(CoordMin.y,CoordMax.y,VarSize);
    C_points(i).Position.z =
unifrnd(CoordMin.z,CoordMax.z,VarSize);
    % Initialize Velocity
    C_points(i).Velocity.x=zeros(VarSize);
    C_points(i).Velocity.y=zeros(VarSize);
    C_points(i).Velocity.z=zeros(VarSize);

    % Evaluation
    [C_points(i).Cost,
C_points(i).Sol]=costfun(C_points(i).Position);
    % Update Personal Best
    C_points(i).Best.Position=C_points(i).Position;
    C_points(i).Best.Cost=C_points(i).Cost;
    C_points(i).Best.Sol=C_points(i).Sol;
    % Update Global Best
    if C_points(i).Best.Cost<GlobalBest.Cost
        GlobalBest=C_points(i).Best;
    end
end
% Array to Hold Best Cost Values at Each Iteration
BestCostpath=zeros(Iter,1);
%% PSO Main Loop
for t=1:Iter
    for i=1:N_pop
        % Update Velocity in x-direction
        C_points(i).Velocity.x =
w*C_points(i).Velocity.x ...
            +
c1*rand(VarSize).*(C_points(i).Best.Position.x-
C_points(i).Position.x) ...
            + c2*rand(VarSize).*(GlobalBest.Position.x-
C_points(i).Position.x);
```

```
        % Update Position in x-direction
        C_points(i).Position.x = C_points(i).Position.x
+ C_points(i).Velocity.x;
        % Check Position of the control points within
x-limits
        C_points(i).Position.x =
max(C_points(i).Position.x,CoordMin.x);
        C_points(i).Position.x =
min(C_points(i).Position.x,CoordMax.x);

        % Update Velocity in y-direction
        C_points(i).Velocity.y =
w*C_points(i).Velocity.y
            +
c1*rand(VarSize).*(C_points(i).Best.Position.y-
C_points(i).Position.y) ...
            + c2*rand(VarSize).*(GlobalBest.Position.y-
C_points(i).Position.y);
        % Update Position in y-direction
        C_points(i).Position.y = C_points(i).Position.y
+ C_points(i).Velocity.y;
        % Check Position of the control points within
y-limits
        C_points(i).Position.y =
max(C_points(i).Position.y,CoordMin.y);
        C_points(i).Position.y =
min(C_points(i).Position.y,CoordMax.y);

        % Update Velocity in z-direction
        C_points(i).Velocity.z =
w*C_points(i).Velocity.z ...
            +
c1*rand(VarSize).*(C_points(i).Best.Position.z-
C_points(i).Position.z) ...
            + c2*rand(VarSize).*(GlobalBest.Position.z-
C_points(i).Position.z);
        % Update Position in z-direction
        C_points(i).Position.z = C_points(i).Position.z
+ C_points(i).Velocity.z;
        % Check Position of the control points within
z-limits
        C_points(i).Position.z =
max(C_points(i).Position.z,CoordMin.z);
        C_points(i).Position.z =
min(C_points(i).Position.z,CoordMax.z);
```

```
        % Evaluation
        [C_points(i).Cost,
C_points(i).Sol]=costfun(C_points(i).Position);
        % Update Personal Best
        if C_points(i).Cost<C_points(i).Best.Cost

C_points(i).Best.Position=C_points(i).Position;
            C_points(i).Best.Cost=C_points(i).Cost;
            C_points(i).Best.Sol=C_points(i).Sol;
            % Update Global Best
            if C_points(i).Best.Cost<GlobalBest.Cost
                GlobalBest=C_points(i).Best;
            end
        end
    end
    % Update Best Cost Ever Found
    BestCostpath(t)=GlobalBest.Cost;
    % Inertia Weight Damping
    w=w*wdamp;
    % Show Iteration Information
    disp(['Iteration ' num2str(t) ': Best Cost = '
num2str(BestCostpath(t))]);
    % Plot Solution
    figure(1);
    PlotPath(GlobalBest.Sol,problem);
    pause(0.01);
end

%% Results
figure;
plot(BestCostpath,'LineWidth',2);
xlabel('Iteration');
ylabel('Best Cost');
grid on;
```

Function that should plot the instantaneous path for each iteration

```
function PlotPath(path,problem)

    xs=problem.xs;
    ys=problem.ys;
    zs=problem.zs;
    xt=problem.xt;
    yt=problem.yt;
    zt=problem.zt;
    xob=problem.xob;
    yob=problem.yob;
    zob=problem.zob;
    rob=problem.rob;

    X=path.X;
    Y=path.Y;
    Z=path.Z;
    xx=path.xx;
    yy=path.yy;
    zz=path.zz;
[s1,s2,s3] = sphere;
    for k=1:numel(xob)
            surf(s1*rob(k)+xob(k), s2*rob(k)+yob(k),
s3*rob(k)+zob(k));
          hold on;
    end
    plot3(xx,yy,zz,'k','LineWidth',2);
    plot3(X,Y,Z,'ro');

plot3(xs,ys,zs,'bs','MarkerSize',12,'MarkerFaceColor','
y');

plot3(xt,yt,zt,'kp','MarkerSize',16,'MarkerFaceColor','
g');
    hold off;
    grid on;
    axis equal;

end
```

6.6 Addition to Collision Detection

In Example 6.2, we proposed a sequential programming procedure to detect the collision between a path and obstacle. Now, let us discuss another aspect for collision detection which is related to the motion planning of the robot while searching for an

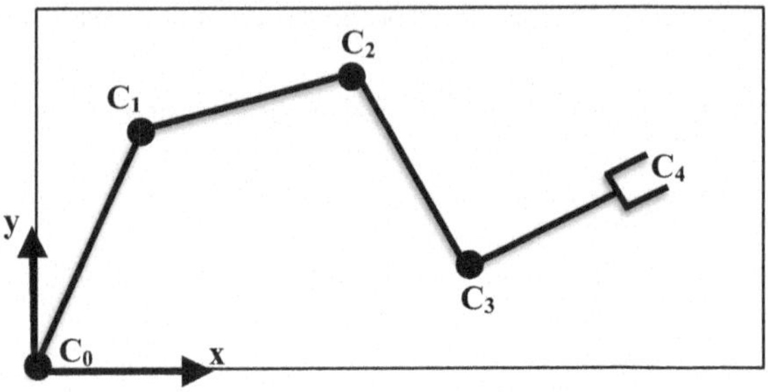

Fig. 6.6 Planar four-link manipulator

inverse kinematic solution for a point in the workspace. Assume we have the 4DOF planar manipulator without rotated end-effector as shown in Fig. 6.6. We can find the position vector of the end-effector C_4 as usual, by taking the first three elements of the fourth column in the overall homogenous transformation matrix H_4^0.

$$H_4^0 = H_1^0 H_2^1 H_3^2 H_4^3, \tag{6.4}$$

In the same way, we can find the position vector at point C_3,

$$H_3^0 = H_1^0 H_2^1 H_3^2, \tag{6.5}$$

and C_2,

$$H_2^0 = H_1^0 H_2^1, \tag{6.6}$$

and C_1 from H_1^0. We already know that C_0 is the base point of the robot $C_0 = \begin{bmatrix} 0 \, 0 \, 0 \end{bmatrix}^T$. We can use the same procedure in Sect. 6.6 to impose the motion of the robot to be in the given bounding box, which represents the working area of the manipulator.

For an inverse kinematic solution that drives at least one of the joints to be in contact with limited working space, we shall add a penalty on that solution. In this way, we can impose the optimization algorithm to search for another feasible solution with a collision-free configuration. Of course, we can check more points on the manipulator, like the centre of masses of the links.

Example 6.3 Recall the 7DOF manipulator in Example 4.4 where it is required to find an inverse kinematic solution considering collision avoidance. Suppose that the working space is as shown in Fig. 6.7, where no part of the robot should touch the bounding box defined by 2500, 1000, and 1500 mm in x, y, and z-direction. This constrained optimization problem is solved with MATLAB code by checking only that the position vector J (see forward function) should not collide with the limits of the configuration space. Vector J represents the position vector of the 6th link in the manipulator. Other points on the robot arm must also be checked to make sure

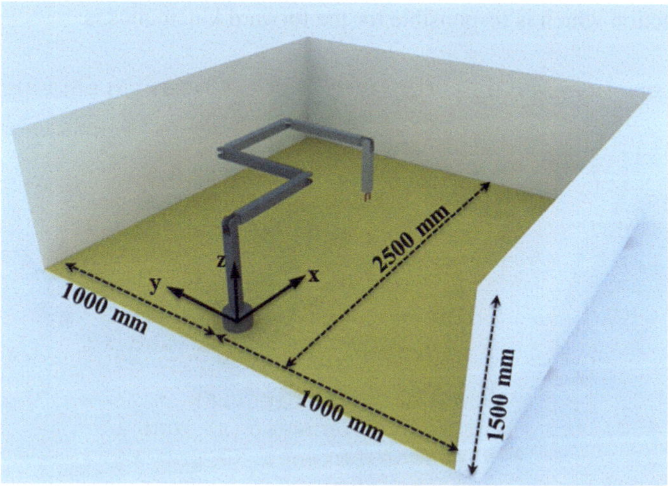

Fig. 6.7 Dimensions of the workspace of the 7DOF manipulator

they do not collide with walls, but it is left for the reader to modify the code as an exercise.

MATLAB code

The cost function that should be optimized is

```
function z = cost(sol)
t1 = sol(1);
t2 = sol(2);
t3 = sol(3);
t4= sol(4);
t5= sol(5);
t6= sol(6);
t7= sol(7);
[x,y,z,J] = forward(t1,t2,t3,t4,t5,t6,t7);
z= sqrt((x-1560)^2+(y)^2+(z-1005)^2);
         if(abs(J.x)>2500 || abs(J.y)>1000||
abs(J.z)>1500)
           z=z+1000;
         end
end
```

The function which is responsible for the forward kinematics is:

```
function [x,y,z,J] = forward(t1,t2,t3,t4,t5,t6,t7)
format short G
th_1= t1;      % joint angle 1
tw_1 = -90;        %twist angle for link 1
a_1 = 0;           %distacne between z1 and z2
d_1 = 1000;        % distance between x1 and x2
th_2= t2;      % joint angle 2
tw_2 = 90;        %twist angle for link 2
a_2 = 600;        %distance between z2 and z3
d_2 = 0;          % distance between x2 and x3
th_3= 90+t3;      % joint angle 3
tw_3 = 0;         %twist angle for link 3
a_3 = 600;        %distance between z3 and z4
d_3 = 0;        % distance between x3 and x4
th_4= t4-90;      % joint angle 4
tw_4 = 0;         %twist angle for link 4
a_4 = 600;        %distance between z4 and z5
d_4 = 0;          % distance between x4 and x5
th_5= t5-90;      % joint angle 5
tw_5 = -90;       %twist angle for link 5
a_5 = 600;        %distance between z5 and z6
d_5 = 0;          % distance between x5 and x6
th_6= t6;      % joint angle 6
tw_6 = -90;       %twist angle for link 6
a_6 = 0;          %distance between z6 and z7
d_6 = 0;          % distance between x6 and x7
th_7= t7;      % joint angle 7
tw_7 = 0;         %twist angle for link 7
a_7 = 0;          %distance between z7 and zn
d_7 = 400;        % distance between x7 and xa
H_1= Denavit(th_1,tw_1,a_1,d_1);
H_2= Denavit(th_2,tw_2,a_2,d_2);
H_3= Denavit(th_3,tw_3,a_3,d_3);
H_4= Denavit(th_4,tw_4,a_4,d_4);
H_5= Denavit(th_5,tw_5,a_5,d_5);
H_6= Denavit(th_6,tw_6,a_6,d_6);
H_7= Denavit(th_7,tw_7,a_7,d_7);
TH = H_1*H_2*H_3*H_4*H_5*H_6*H_7;
x=TH(1,4);
y=TH(2,4);
z=TH(3,4);
C = H_1*H_2*H_3*H_4*H_5*H_6;
J.x = C(1,4);
J.y = C(2,4);
J.z = C(3,4);
end
```

The function that returns a homogenous transformation matrix for each link is:

```
function [DH] = Denavit(theta,alpha,a,d)
DH = [cosd(theta),-
1*sind(theta)*cosd(alpha),sind(theta)*sind(alpha),a*cos
d(theta);...
    sind(theta),cosd(theta)*cosd(alpha),-
1*cosd(theta)*sind(alpha),a*sind(theta);...
    0,sind(alpha),cosd(alpha),d;...
    0,0,0,1];
end
```

The particle swarm optimization (PSO) main program is

```
clc;
clear;
close all;
%% problem parameters
costfun = @(x) cost(x);
nvar = 7;
V_size = [1,nvar];
VarMin= [-180,-180,-180,-180,-180,-180,-360];           %
Decision Variables Lower Bound
VarMax= [180,180,180,180,180,180,360];           %
Decision Variables Upper Bound
%% PSO parameters
nIter = 200;
nPop = 100;
w = 1;
damping = 0.95;
c1 = 1; % personal acceleration
c2 = 1; % social acceleration
%% initialization
particle_templet.position = [];
particle_templet.velocity = [];
particle_templet.cost = [];
particle_templet.best.position = [];
particle_templet.best.cost = [];
% create particles
particle = repmat(particle_templet, 50,1);
%
Globalbest.cost = Inf; % start with the worst value
% initialize population
for i=1:nPop
%     particle(i).position =
unifrnd(VarMin,VarMax,V_size);
```

```
    for j=1:length(VarMin)
      particle(i).position(j) =
unifrnd(VarMin(j),VarMax(j),1) ;
    end
    particle(i).velocity = zeros(V_size);
    particle(i).cost = costfun(particle(i).position);
    % update personal best
    particle(i).best.position = particle(i).position;
     particle(i).best.cost = particle(i).cost;
     % update global best
     if (particle(i).best.cost < Globalbest.cost)
         Globalbest = particle(i).best;
     end
end
% array to hold best costs value for each iteration
BestCost_Value = zeros(nIter,1);


%% main loop
for it =1:nIter

    for i=1:nPop
        particle(i).velocity =
w*particle(i).velocity+...
            c1*rand(V_size).*(particle(i).best.position
- particle(i).position)+...
            c2*rand(V_size).*(Globalbest.position -
particle(i).position);
        particle(i).position = particle(i).position +
particle(i).velocity;
        for j=1:length(VarMin)
            particle(i).position(j) =
max(particle(i).position(j), VarMin(j));
            particle(i).position(j) =
min(particle(i).position(j), VarMax(j));
        end
        % update personal best
        particle(i).cost =
costfun(particle(i).position);
        if(particle(i).cost < particle(i).best.cost)
            particle(i).best.position =
particle(i).position;
            particle(i).best.cost = particle(i).cost;
            % update global best
            if(particle(i).best.cost < Globalbest.cost)
                Globalbest = particle(i).best;
```

```
                end

            end

        end
        BestCost_Value(it) = Globalbest.cost;
        disp(['iteration' num2str(it) ': Best cost = '
    num2str(BestCost_Value(it))]);
        w = w*damping;
    end
```

6.7 Trajectory Planning

We have studied the motion of the robot and developed mathematical expressions for the velocity and acceleration kinematics. All of the driven expressions for Jacobian and velocities were as a function of the joint variables. In real life, robots are actuated by motors, hydraulic systems, pneumatic systems, etc., and all of these actuation systems should work with respect to a time variable. For instance, we have to control how much time we need to operate a motor or press a hydraulic cylinder.

We have also seen how to plan a path that the robot should follow in case of a mobile robot or only the end-effector in case of manipulators. Again, path planning is just a geometric description of the motion without any information about *how* this motion occurs. Trajectory planning enables us to describe the velocity and acceleration along the path, and it is the operation that we need to map representation of the motion from joint variables domain to the time domain. In other words, we have to derive equations for the position, velocity, and acceleration as a function of time.

$$q = q(t),$$
$$\dot{q} = \dot{q}(t),$$
$$\ddot{q} = \ddot{q}(t). \tag{6.7}$$

The joint actuator is the physical source of the motion, and controlling angular position and angular velocity of that actuator leads to control velocity and acceleration of any point on the robotic arm. Thus, Cartesian points on the path should be represented in the joint space of the robot. As we have studied before, inverse kinematics is the operation that maps a point from Cartesian space to the joint space. Consequently, all the points on the path must be represented in the joint space using an inverse kinematic solution.

A path can be defined as a continuous function $q(t) : \left[t_0 \; t_f \right]$ with $q(t_0) = q_{\text{initial}}$ and $q(t_f) = q_{\text{final}}$. Figure 6.8 shows the boundary conditions, which are initial and final configurations as well as via points. A via point provides extra information on the shape of the trajectory function. The number of via points depends on the topology of

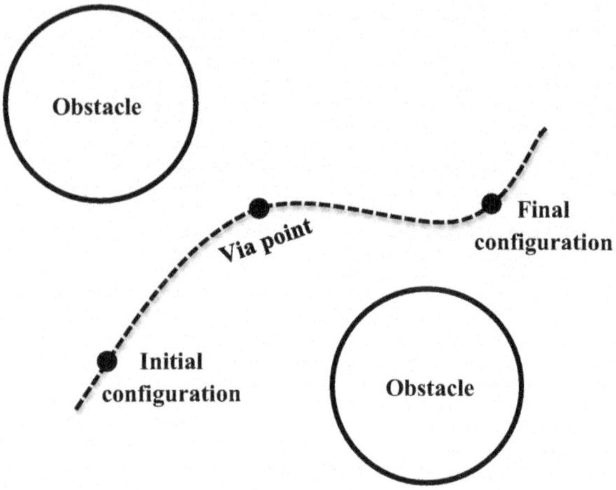

Fig. 6.8 Boundary condition and via point on a path

the workspace with its obstacles, so there is no specific number for all applications. $q(t)$ holds the specifications of the trajectory from the initial time to the end of the run-time. Consequently, by taking the first derivative, we find the velocity equation, while by taking the second derivative, we will find the acceleration equation.

We can see that the initial and final configurations are always given, while via points are obtained from path planning. In other words, path planning provides us with the optimum via points necessary for trajectory planning. In industry, there are other ways to generate via points, like manually driving the end-effector to the desired points or by using touch pendent. By recording these joint variables, we can run the robot in an operation called teach and playback. For unconstrained motion, the robot can move from one configuration to another and there is no need for via points, since there are no obstacles. A common technique for trajectory planning is to use a polynomial and apply the boundary conditions to calculate all the required coefficients. For each joint of the manipulator, a trajectory function should be driven. Thus, the number of the trajectories is equal to the degree of the freedom of the robot manipulator. This means that for each joint variable, there are angular position equations that can be derived once to find the angular velocity equation and derived again to find the angular acceleration equation of the joint variable.

6.7.1 Point-to-Point Trajectory

In this section, we will consider the case where only the initial configuration and final configuration of the manipulator are given. We will suppose that a polynomial of the third degree can represent the trajectory to ensure that there is a second derivative for the polynomial. If we are considering the derivative of the acceleration, which is called a jerk, then we shall assume a polynomial of the fourth degree to represent the trajectory. We will assume that the following polynomial for a specific joint variable

$$q(t) = a_0 + a_1 t + a_2 t^2 + a_3 t^3. \tag{6.8}$$

Velocity equation and acceleration equation can be obtained by taking the first and second derivative of Eq. (6.8), respectively:

$$\dot{q}(t) = a_1 + 2a_2 t + 3a_3 t^2, \tag{6.9}$$

$$\ddot{q}(t) = 2a_2 + 6a_3 t. \tag{6.10}$$

The solution for Eq. (6.8) is found by calculating the vector of the coefficients $\begin{bmatrix} a_0 \; a_1 \; a_2 \; a_3 \end{bmatrix}^T$, and this can be done using boundary conditions. Initially, the quantities at time equal to t_0 are

$$q(t_0) = q_0, \tag{6.11}$$

$$\dot{q}(t_0) = v_0, \tag{6.12}$$

$$\ddot{q}(t_0) = \alpha_0, \tag{6.13}$$

while at the final configuration at a time equal to t_f, they are

$$q(t_f) = q_f, \tag{6.14}$$

$$\dot{q}(t_f) = v_f, \tag{6.15}$$

$$\ddot{q}(t_f) = \alpha_f. \tag{6.16}$$

By substituting Eqs. (6.11), (6.12), (6.14), and (6.15) in Eqs. (6.8) and (6.9), we can find the coefficients from the following equation:

$$\begin{bmatrix} 1 & t_0 & t_0^2 & t_0^3 \\ 0 & 1 & 2t_0 & 3t_0^2 \\ 1 & t_f & t_f^2 & t_f^3 \\ 0 & 1 & 2t_f & 3t_f^2 \end{bmatrix} \begin{bmatrix} a_0 \\ a_1 \\ a_2 \\ a_3 \end{bmatrix} = \begin{bmatrix} q_0 \\ v_0 \\ q_f \\ v_f \end{bmatrix}. \tag{6.17}$$

The coefficients in Eq. (6.17) can be obtained if we know the values of the initial and final position and velocity. Equation (6.17) can be written as

$$t\,a = b, \tag{6.18}$$

$$a = t^{-1} b. \tag{6.19}$$

In this case, we have to avoid using boundary conditions in (6.13) and (6.16) to ensure that t will be a 4×4 square matrix. If we include acceleration values from (6.13) and (6.16), then matrix t will be 6×4 nonsquare and cannot be inverted. To include acceleration states, we may extend the polynomial to the 5th order to ensure that we will have a 6×6 square t matrix that can be inverted.

6.7.2 Via-Point Trajectory

Via-point trajectories coincide with the original geometric path, depending on the number of the points themselves. A point-to-point trajectory is appropriate for a free workspace, while a via-point trajectory is appropriate for a constrained workspace. These intermediate configurations provide more information about the trajectory between initial configuration and target configuration that leads to a successful avoidance of obstacles. The number of the constraints on the trajectory increases by the number of via points given. For instance, suppose that we have the initial configuration at time t_0, one via point at time t_1, and the final configuration at time t_f and then the constraints on the polynomial will be

$$\begin{aligned} q(t_0) &= q_0 \\ \dot{q}(t_0) &= v_0 \\ \ddot{q}(t_0) &= \alpha_0 \\ q(t_1) &= q_1 \\ q(t_f) &= q_f \\ \dot{q}(t_f) &= v_f \\ \ddot{q}(t_f) &= \alpha_f \end{aligned}$$

If we would like to use all seven constraints on a polynomial, then this polynomial should be of the 6th order:

$$q(t) = a_0 + a_1 t + a_2 t^2 + a_3 t^3 + a_4 t^4 + a_5 t^5 + a_6 t^6. \tag{6.20}$$

As a general rule, we can say that the order of the polynomial should be

$$\text{polynomial order} = \text{Number of the used constraints} - 1. \tag{6.21}$$

Example 6.4 Figure 6.9 shows a path planned using particle swarm optimization given an initial point and target point. Cubic spline has been used for this problem, where three control points give the shape of the cubic spline function and they are the via points for the trajectory planning. The planar 2DOF revolute-joint manipulator shown in Fig. 6.10 is required to follow this path with its end-effector, as shown in Fig. 6.11.

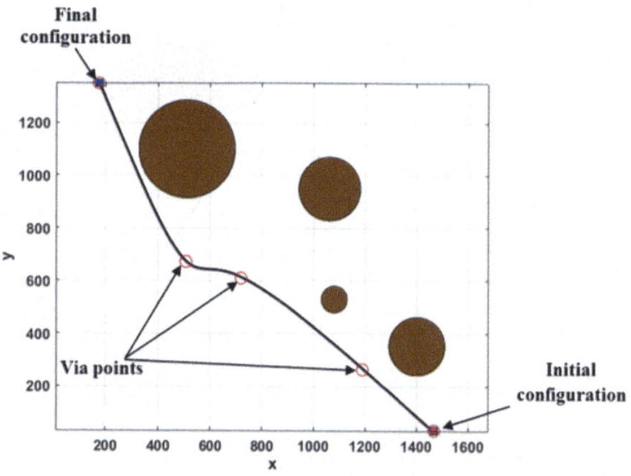

Fig. 6.9 Via points on a given path

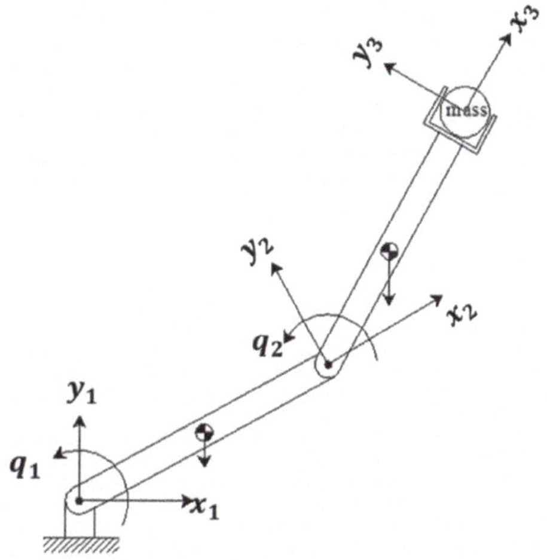

Fig. 6.10 2DOF planar revolute-joint manipulator

Table 6.1 shows the start, target, and via points in the Cartesian space and mapped to the joint space using inverse kinematics.

For this example, we will assume that the robot begins to move from the start point at time equal to zero $t_0 = 0$ and reaches the final configuration at time $t_f = 1$ s. The time interval between 0 and 1 s is equally spaced so that robot reaches configuration 1 at time 0.25 s, configuration 2 at time 0.5 s, and 0.75 for the configuration 3. The angular position of the joints is given at start and target points, and we shall assume that the robot moves from the rest of the points and stops only at the final configuration. Consequently, the velocity and acceleration at start and goal points are zero. In addition to the six constraints on the start point and target point, there are also three more constraints, which are the angular positions at configurations 1, 2, and 3. Thus, we have nine constraints which means that the polynomial should be of the 8th order to represent the trajectory

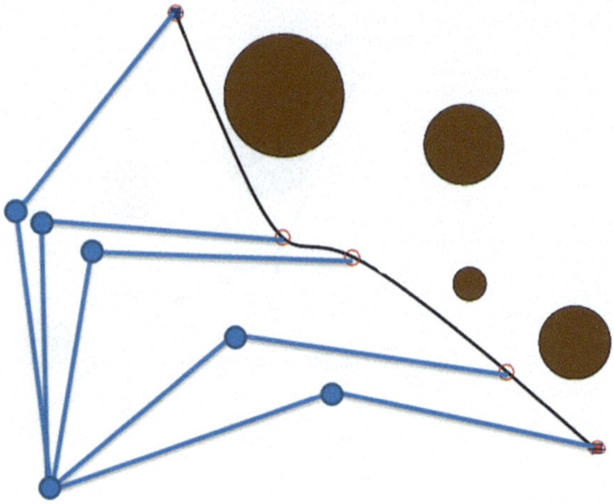

Fig. 6.11 End-effector of the planar robot follows a given path

Table 6.1 Configurations on the given path

Point	x-axis mm	y-axis mm	Corresponding joint angles degree	
			q_1	q_2
1	1191	266.9	48.2	−71.1
2	721.4	612.6	91.2	−101.7
3	504.5	674.1	109	−111.7
Start point	1470	33	12.7	22.82
Goal point	174	1350	107.5	−49.7

$$q(t) = a_0 + a_1 t + a_2 t^2 + a_3 t^3 + a_4 t^4 + a_5 t^5 + a_6 t^6 + a_7 t^7 + a_8 t^8 \tag{6.22}$$

The boundary conditions are
For joint variable 1:

$$q(0) = 12.7,\ \dot{q}(0) = 0,\ \ddot{q}(0) = 0,\ q(1) = 107.5,\ \dot{q}(1) = 0,\ \ddot{q}(1) = 0,$$
$$q(0.25) = 48.2,\ q(0.5) = 91.2,\ q(0.75) = 109$$

For joint variable 2:

$$q(0) = 22.82,\ \dot{q}(0) = 0,\ \ddot{q}(0) = 0,\ q(1) = -49.7,\ \dot{q}(1) = 0,\ \ddot{q}(1) = 0,$$
$$q(0.25) = -71.1,\ q(0.5) = -101.7,\ q(0.75) = -111.7$$

Applying the above boundary conditions on Eq. (6.22) yields:

$$q_1(t) = 12.7 + 8383.4t^3 - 41147t^4 + 88190t^6 - 98866t^7 + 56593t^8 \quad (6.23)$$

$$q_2(t) = 22.82 - 33651t^3 + 2.0385 \times 10^5 t^4 - 5.1 \times 10^5 t^6 + 6.41 \times 10^5 t^7 - 4 \times 10^5 t^8. \quad (6.24)$$

MATLAB Code

```
clear;

clc;

start_time = 0;
point1_time  = 0.25;

point2_time  = 0.5;

point3_time  = 0.75;

end_time  = 1;

joint1_s = [12.7,0,0];

joint1_1 = 48.2;

joint1_2 = 91.2;

joint1_3 = 109;

joint1_g = [107.5,0,0];

%%%%%%%%

joint2_s = [22.82,0,0];

joint2_1 = -71.1;

joint2_2 = -101.7;

joint2_3 = -111.7;

joint2_g = [-49.7,0,0];
```

```
para =
[pos(start_time);velo(start_time);accel(start_time);...

     pos(point1_time);...

     pos(point2_time);...

     pos(point3_time);...

     pos(end_time);velo(end_time);accel(end_time)];


b = [joint1_s,joint1_1,joint1_2,joint1_3,joint1_g]';
a = para\b;

q_j1 = @(t)
a(1)+a(2)*t+a(3)*t^2+a(4)*t^3+a(5)*t^4+a(6)*t^5+a(7)*t^
6+a(8)*t^7+a(9)*t^8; % angular position joint1

v_j1 = @(t)
a(2)+2*a(3)*t+3*a(4)*t^2+4*a(5)*t^3+5*a(6)*t^4+6*a(7)*t
^5+7*a(8)*t^6+8*a(9)*t^7; % angular velocity joint1
a_j1 = @(t)
2*a(3)+6*a(4)*t+12*a(5)*t^2+20*a(6)*t^3+30*a(7)*t^4+42*
a(8)*t^5+56*a(9)*t^6; % angular acceleration joint1

%%%%%%%%%%%%%%%%%%%

c = [joint2_s,joint2_1,joint2_2,joint2_3,joint2_g]';

k = para\c;

q_j2 = @(t)
k(1)+k(2)*t+k(3)*t^2+k(4)*t^3+k(5)*t^4+k(6)*t^5+k(7)*t^
6+k(8)*t^7+k(9)*t^8; % angular position joint2

v_j2 = @(t)
k(2)+2*k(3)*t+3*k(4)*t^2+4*k(5)*t^3+5*k(6)*t^4+6*k(7)*t
^5+7*k(8)*t^6+8*k(9)*t^7; % angular velocity joint2

a_j2 = @(t)
2*k(3)+6*k(4)*t+12*k(5)*t^2+20*k(6)*t^3+30*k(7)*t^4+42*
k(8)*t^5+56*k(9)*t^6; % angular acceleration joint2
```

The results of the angular position of the joint 1 and joint 2 are shown in Fig. 6.12. Figure 6.13 illustrates the velocity equation, and Fig. 6.14 reveals the acceleration

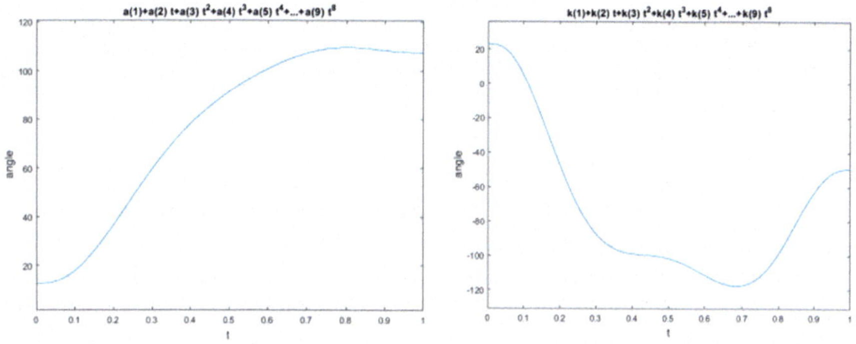

Fig. 6.12 Joint angles on the trajectory

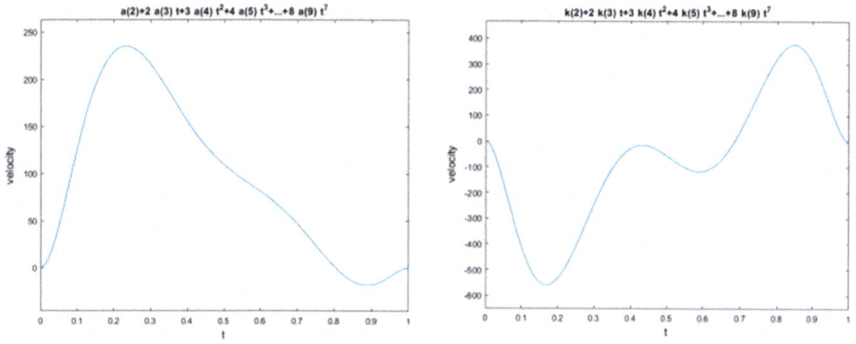

Fig. 6.13 Joint velocities on the trajectory

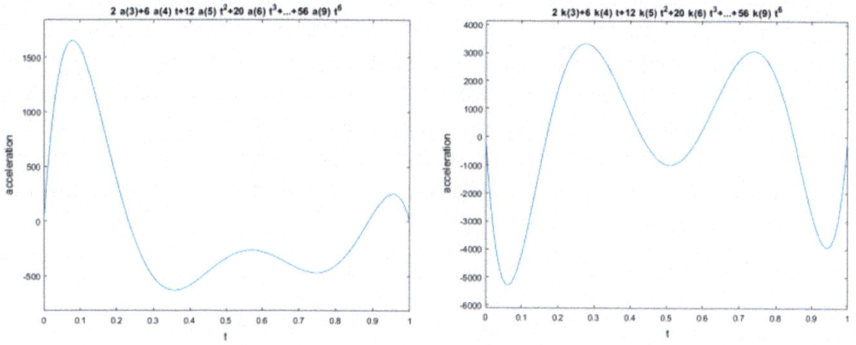

Fig. 6.14 Joint accelerations on the trajectory

values on the trajectory.

6.8 Summary

In this chapter, we have learned how to find the shortest path between two points in a constrained environment. Path planning returns a geometric graph which is a set of Cartesian points and has no information about the motion on that path. Consequently, we have introduced the concept of trajectory planning to describe the velocity and acceleration of joint variables during motion of the robot on the Cartesian path. Trajectory planning occurs in joint space, and thus, we have used inverse kinematics to map the Cartesian points to the joint space. The velocity and acceleration values are essential parameters to drive the dynamic equations in chapter seven.

Exercises

6.1 In Sect. 6.6, we have described how to avoid a circular obstacle in detail. Modify Example 6.2 considering rectangular obstacles instead of circular ones.

6.2 Use ant colony optimization to find the shortest path between two points by dividing the workspace into a grid of equally spaced elements.

6.3 Modify the code of Example 6.1 to solve the classical travelling salesman problem. Use the same given points.

6.4 In Example 6.3, we have checked only one joint to make sure that it does not collide with the limits of the workspace. Consider checking all the joints of the robot manipulator for collision detection, as well as the centre of each link.

6.5 We have derived Eq. (6.17) by ignoring the acceleration constraints. Derive the same equation by considering acceleration values at the start and final configurations and propose an appropriate polynomial to represent the trajectory.

6.6 Write a MATLAB code for the derived equation in Exercise 5 and assume that the robot starts, moves through the rest of the points, and reaches its final configuration at time = 2 s.

References

Brand M, Masuda M, Wehner N, Yu X-H (2010) Ant colony optimization algorithm for robot path planning. In: 2010 international conference on computer design and applications, 2010. IEEE, pp V3–436-V433–440

Cheney EW, Kincaid DR (2012) Numerical mathematics and computing. Cengage Learning

Raja P, Pugazhenthi S (2012) Optimal path planning of mobile robots: a review. Int J Phys Sci 7(9):1314–1320

Chapter 7
Dynamics

7.1 Introduction

The kinematic equations studied in the previous chapters do not describe the relationship between the motion and its producers, i.e. forces and torques. In this chapter, we deal with dynamic equations to take into account that relationship. The dynamic equations are fundamental, especially when we need to specify the specifications of the actuators of the robot. Also, dynamic equations are important for designing control systems, simulation, mechanical design, and animation of the robot. We will introduce the principle of the Euler–Lagrange equations to drive dynamic equations. Also, the symbolic toolbox in MATLAB will be used as an efficient way to drive the equations for manipulators with high degrees of freedom, where this operation is very tedious to calculate by hand.

7.2 Euler–Lagrange Equations

Euler–Lagrange equations describe the dynamic equations of a system, and they correspond to equations derived from Newton's second law. However, the Euler–Lagrange equations are preferable, especially for complex mechanical systems like multilinkage mechanisms and robots.

Assume that we have a point of mass moving in the positive direction of y-axis due to the application of an external force f and gravitational force mg downwards, as shown in Fig. 7.1. By Newton's second law, we can say that

$$m\,\ddot{y} = f - mg \tag{7.1}$$

The term $m\,\ddot{y}$ in Eq. (7.1) can be written as

H. N. Ghafil and K. Jármai (ed.), *Optimization for Robot Modelling with MATLAB*,
https://doi.org/10.1007/978-3-030-40410-9_7

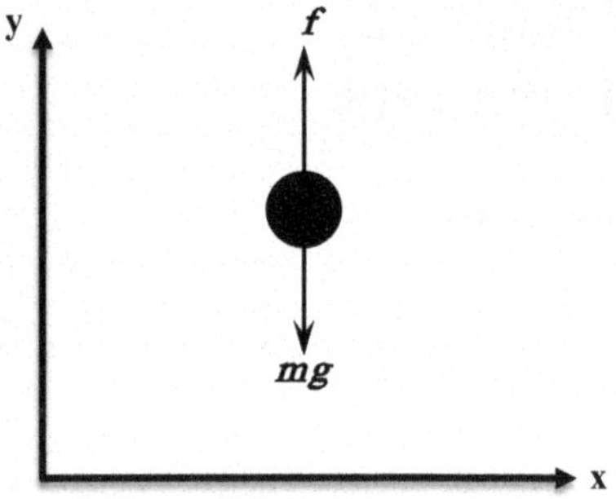

Fig. 7.1 Forces subjected to the point of mass moving in one dimension

$$m\,\ddot{y} = \frac{\mathrm{d}}{\mathrm{d}t}(m\,\dot{y}) = \frac{\mathrm{d}}{\mathrm{d}t}\frac{\partial}{\partial \dot{y}}\left(\frac{1}{2}m\,\dot{y}^2\right) = \frac{\mathrm{d}}{\mathrm{d}t}\frac{\partial K}{\partial \dot{y}}, \tag{7.2}$$

where $K = \frac{1}{2}m\,\dot{y}^2$ is the kinetic energy. In the same way, the term $m\,g$ can be written as

$$m\,g = \frac{\partial}{\partial y}(mgy) = \frac{\partial P}{\partial y}, \tag{7.3}$$

where $P = mgy$ is the potential energy.

The Lagrangian L of a system is defined as the difference between kinetic and potential energy,

$$L = K - P = \frac{1}{2}m\dot{y}^2 - mgy. \tag{7.4}$$

Differentiation of Eq. (7.4) with respect to $\dot{y}$ and y yields

$$\frac{\partial L}{\partial \dot{y}} = m\dot{y} = \frac{\partial K}{\partial \dot{y}}, \tag{7.5}$$

$$\frac{\partial L}{\partial y} = -mg = -\frac{\partial P}{\partial y}. \tag{7.6}$$

Substitute Eq. (7.5) into (7.2)

$$m\ddot{y} = \frac{\mathrm{d}}{\mathrm{d}t}\left(\frac{\partial K}{\partial \dot{y}}\right). \tag{7.7}$$

Also, substitute (7.6) into (7.3)

$$m\,g = -\frac{\partial L}{\partial y}. \tag{7.8}$$

Now, substituting the values of Eqs. (7.7) and (7.8) into Eq. (7.1) yields

$$\frac{\mathrm{d}}{\mathrm{d}t}\left(\frac{\partial K}{\partial \dot{y}}\right)-\frac{\partial L}{\partial y}=f \tag{7.9}$$

Equation (7.9) is called the Euler–Lagrange equation, and it describes the motion of a body subjected to generalized forces f. Generalized forces represent all the forces and torques exerted on a body, like actuation and damping forces that cannot be derived from potential function.

7.3 Rigid Links

In Sect. 7.2, we have derived Euler–Lagrange equations for a body with a single degree of freedom. This is appropriate when we would like to find the dynamic equation of a rigid link with a prismatic joint, since the whole link can be represented by its centre of mass, as shown in Fig. 7.2.

The term rigid link implies that the potential energy can be driven only from the differences of the heights from a reference line. Thus, in this book we will not consider the potential energy caused by the elastic behaviour of the body. Let us consider deriving the Euler–Lagrange equation for a link with revolute joints, such as in Fig. 7.3. The rest or reference position of the link is along the dashed line in Fig. 7.4; at that position, there is no potential energy. Any movement in any direction

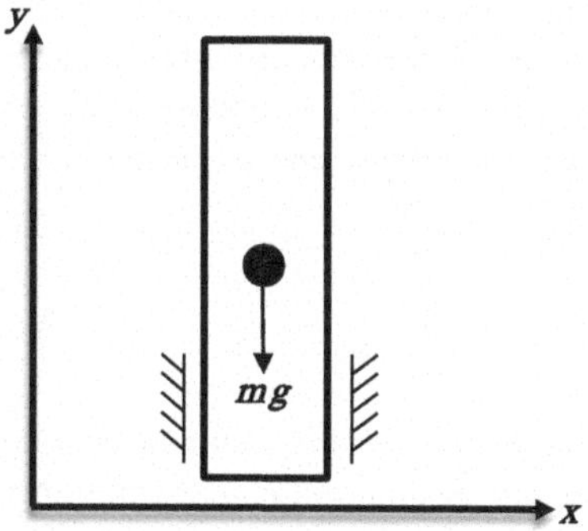

Fig. 7.2 Rigid link with prismatic joint

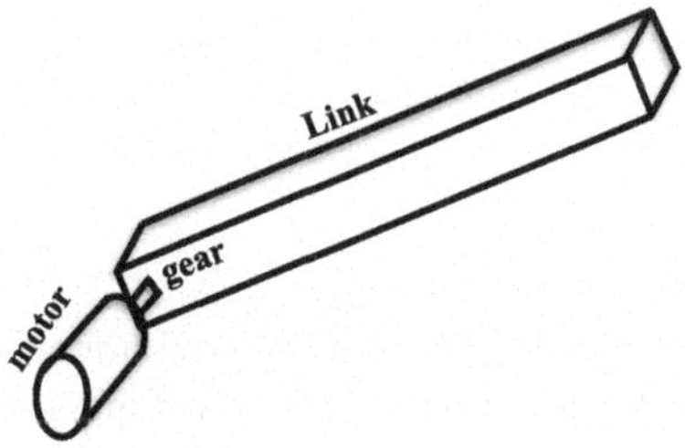

Fig. 7.3 Link connected to a motor via a gear

Fig. 7.4 Description of the potential energy of a rotated link

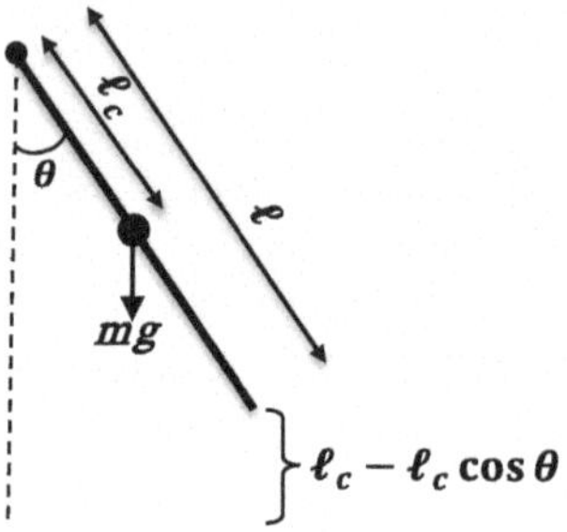

causes the head to rise by $\ell_c(1 - \cos(\theta))$. Thus, the potential energy will be

$$P = mg\ell_c(1 - \cos\theta_\ell) \tag{7.10}$$

where ℓ_c is the distance between the axis of rotation and centre of mass of the link and θ_ℓ denotes the rotation angle of the link.

The kinetic energy is a composite of the rotational energy of the link plus rotational energy of the motor engaged to the link. We can assume that the kinetic energy of the link K_ℓ has both translational and rotational parts

$$K_\ell = \frac{1}{2} II_{c.m}\dot{\theta}^2 + \frac{1}{2} m{\upsilon_{c.m}}^2, \tag{7.11}$$

where $II_{c.m}$ and $\upsilon_{c.m}$ are the moment of inertia expressed at the centre of mass of the link and linear velocity of the centre of mass, and $\upsilon_{c.m} = \ell_c\dot{\theta}_\ell$. We can get the same result if we assume that the link is entirely rotated about the pivot (joint), and in this case, the kinetic energy of the link is only rotational energy without a translational part

$$K_\ell = \frac{1}{2} II_{\text{joint}}\dot{\theta}^2, \tag{7.12}$$

where II_{joint} is the moment of inertia about the axis of rotation of the joint. The second assumption is more appropriate in our task in this chapter. Thus, the kinetic energy of the system is the summation of the rotational energy of the link about its rotation axis plus the rotational energy of the motor.

$$K = \frac{1}{2} II_m\dot{\theta}_m^2 + \frac{1}{2} II_\ell\dot{\theta}_\ell^2 \tag{7.13}$$

$$\theta_m = r\theta_\ell, \tag{7.14}$$

where II_m, II_ℓ, and r are rotational inertia of the motor, rotational inertia of the link, and gear ratio, respectively. Equation (7.13) can be rewritten as

$$K = \frac{1}{2}(r^2 II_m + II_\ell)\dot{\theta}_\ell^2 \tag{7.15}$$

By assuming that $J = r^2 J_m + J_\ell$, we can write

$$K = \frac{1}{2} II\dot{\theta}_\ell^2 \tag{7.16}$$

Equation (7.16) is appropriate for use with a link with revolute joints and actuated by motors because it represents the kinetic energy in terms of the rotation of the link and the overall inertia *J*, which include both the inertia of the motor and link.

The Lagrangian of the system will be

$$L = \frac{1}{2} J\dot{\theta}_\ell^2 + mg\ell_c(1 - \cos\theta_\ell) \tag{7.17}$$

By using the expression in Eq. (7.9), we get

$$J\ddot{\theta}_\ell - mg\ell_c \sin\theta_\ell = \tau_\ell. \tag{7.18}$$

τ_ℓ is the net torque that reflected the link from the motor and is equal to the torque of the motor τ_m minus damping torques in the motor and links $B_m\dot{\theta}_m$ and $B_\ell\dot{\theta}_\ell$, respectively:

$$\tau_\ell = \tau_m - B_m\dot{\theta}_m - B_\ell\dot{\theta}_\ell. \tag{7.19}$$

Arrangement of Eq. (7.19) yields

$$\tau_\ell = \tau_m - B\dot{\theta}_\ell \tag{7.20}$$

B is the overall damping coefficient and equal to $B = rB_m + B_\ell$ where B_m and B_ℓ are damping coefficients of the motor and link, respectively. Substituting (7.20) into (7.18) yields

$$J\ddot{\theta}_\hbar + B\dot{\theta}_\ell - mg\ell_c \sin\theta_\ell = \tau_m \tag{7.21}$$

Equation (7.21) represents the equation of motion of the system in terms of the damping torques and motor torque, and that is our goal: to find the proper specifications of the motor(s) to the manipulator.

7.4 Dynamic Equations for *n*-DOF Manipulator

The kinetic and potential energies of *n*-link robot manipulator are the summation of these energies, and according to that the Lagrangian of the system will be

$$L = \sum_{i=1}^{m} K_i + P_i \tag{7.22}$$

The kinetic energy has two parts: rotational energy and translational energy represented by the concentrated mass at the centre of the mass of the body. It is generally expressed as

$$K = \frac{1}{2}mv^2 + \frac{1}{2}II\omega^2, \tag{7.23}$$

where m, v, I, and w are the mass of the body, linear velocity, moment of inertia about axis of rotation, and angular velocity of the rigid body, respectively. Equation (7.23) can be written in the matrix form

$$K = \sum_{i=1}^{m} \left[\frac{1}{2}mv^T v + \frac{1}{2}\omega^T II\omega \right]. \tag{7.24}$$

We know that each link is attached rigidly to a coordinate frame that the link rotates about. Consequently, the inertia tensor II of a link should be represented to the base frame to be consistent with other parameters in Eq. (7.24)

$$II = RIR^T \tag{7.25}$$

where II is the inertia tensor represented to the base or reference frame while I is a 3×3 inertia tensor represented to the centre of mass of the body.

Thus, kinetic energy can be written as

$$K = \frac{1}{2}mv^T v + \frac{1}{2}\omega^T RIR^T \omega, \tag{7.26}$$

recalling Eqs. (5.36) and (5.37)

$$v = J_v \dot{q}, \quad \omega = J_\omega \dot{q}, \tag{7.27}$$

by substituting Eqs. (7.27) and (7.26) into (7.24), we get

$$K = \sum_{i=1}^{m} \frac{1}{2}\dot{q}^T \left[m_i {J_v}^T J_v + {J_\omega}^T RIR^T J_\omega \right] \dot{q}. \tag{7.28}$$

Example 7.1 Consider the planar manipulator in Example 6.4, where it is required to find the dynamic equations of the robot which holds mass M in its end-effector. Table 7.1 illustrates the link parameters of the manipulator, where each link is 750 mm in length. This example with all numerical data and formulations was presented by

Table 7.1 Spatial parameters of the 2DOF planar manipulator

Link	a_i mm	$\propto_i$	d_i mm	q_i
1	750	0	0	q_1
2	750	0	0	q_2

Ghafil and Jármai (2019), and here we just focus on the formulation of the dynamic equations.

Firstly, path planning has been done through the workspace shown in Fig. 6.8 to find optimum positions for the via points on the path. The start point, goal point, and via points are mapped to the joint space using inverse kinematics for the values shown in Table 6.1. With nine constraints, we used an eighth-order polynomial in Eq. (6.22) for the trajectory planning. Equations (6.23) and (6.24) are the angular position equations for the two joints, which are easily differentiated to find the velocity and acceleration values from time 0 to 1 s.

We should calculate the homogenous transformation matrices at the centre of each link as well as at the end-effector, where the robot is holding a weight of mass (M) with moment of inertia (I_m). Tc_1, Tc_2, T_1, and T_2 represent the homogenous transformation matrix (HTM) at the centre of link 1, centre of link 2, tip of the link 1, and the end-effector, respectively.

$$Tc_1 = \begin{bmatrix} \cos(q_1) & -\sin(q_1) & 0 & lc_{1*}\cos(q_1) \\ \sin(q_1) & \cos(q_1) & 0 & lc_{1*}\sin(q_1) \\ 0 & 0 & 1 & 0 \\ 0 & 0 & 0 & 1 \end{bmatrix}, \tag{7.29}$$

$$Tc_2 = \begin{bmatrix} \cos(q_2) & -\sin(q_2) & 0 & lc_{2*}\cos(q_2) \\ \sin(q_2) & \cos(q_2) & 0 & lc_{2*}\sin(q_2) \\ 0 & 0 & 1 & 0 \\ 0 & 0 & 0 & 1 \end{bmatrix}, \tag{7.30}$$

$$T_1 = \begin{bmatrix} \cos(q_1) & -\sin(q_1) & 0 & l_{1*}\cos(q_1) \\ \sin(q_1) & \cos(q_1) & 0 & l_{1*}\sin(q_1) \\ 0 & 0 & 1 & 0 \\ 0 & 0 & 0 & 1 \end{bmatrix}, \tag{7.31}$$

$$T_2 = \begin{bmatrix} \cos(q_2) & -\sin(q_2) & 0 & l_{2*}\cos(q_2) \\ \sin(q_2) & \cos(q_2) & 0 & l_{2*}\sin(q_2) \\ 0 & 0 & 1 & 0 \\ 0 & 0 & 0 & 1 \end{bmatrix}. \tag{7.32}$$

The linear Jacobian at the centre of the mass of link 1 is

$$J_{vc1} = \frac{\partial o_{c1}^1}{\partial q_i}, \tag{7.33}$$

where $i = 1, 2$ and o^1_{c1} is the position vector of Tc_1.

Also, the angular Jacobian J_w is just the z-vector at the rotation matrix in the HTM,

$$J_{wc1} = \begin{bmatrix} z_1 \; 0 \end{bmatrix} = \begin{bmatrix} 0\,0 \\ 0\,0 \\ 1\,0 \end{bmatrix}, \tag{7.34}$$

$$J_{vc1} = \begin{bmatrix} -l_{c1} * \sin(q_1) & 0 \\ l_{c1} * \cos(q_1) & 0 \\ 0 & 0 \end{bmatrix}, \tag{7.35}$$

$$J_{vc2} = \frac{\partial o^1_{c2}}{\partial q_i}, \tag{7.36}$$

where o^1_{c2} is the position vector of $T_1 x T c_2$.

$$j_{wc2} = \begin{bmatrix} z_1 \; z_2 \end{bmatrix}, \tag{7.37}$$

where z_2 is z-vector of $T_1 x T c_2$

$$J_{vc2} = \begin{bmatrix} -l_1 * \sin(q_1) - l_{c2} * \sin(q_1 + q_2) & -l_{c2} * \sin(q_1 + q_2) \\ l_1 * \cos(q_1) + l_{c2} * \cos(q_1 + q_2) & l_{c2} * \cos(q_1 + q_2) \\ 0 & 0 \end{bmatrix}. \tag{7.38}$$

$$J_{wc2} = \begin{bmatrix} 0\,0 \\ 0\,0 \\ 1\,1 \end{bmatrix}. \tag{7.39}$$

$$J_{v2} = \frac{\partial o^1_2}{\partial q_i}, \tag{7.40}$$

o^1_2 is the position vector of $T_1 * T_2$.

$$J_{w2} = \begin{bmatrix} z_1 \; z_2 \end{bmatrix}, \tag{7.41}$$

$$J_{v2} = \begin{bmatrix} -l_1 * \sin(q_1) - l_2 * \sin(q_1 + q_2) & -l_2 * \sin(q_1 + q_2) \\ l_1 * \cos(q_1) + l_2 * \cos(q_1 + q_2) & l_2 * \cos(q_1 + q_2) \\ 0 & 0 \end{bmatrix}, \tag{7.42}$$

$$J_{w2} = \begin{bmatrix} 0\,0 \\ 0\,0 \\ 1\,1 \end{bmatrix}, \tag{7.43}$$

The translational part of the kinetic energy is:

$$k_v = \frac{1}{2}\dot{q}^T\left\{m_1 J_{vc1}^T J_{vc1} + m_2 J_{vc2}^T J_{vc2} + M J_{v2}^T J_{v2}\right\}\dot{q}, \tag{7.44}$$

Substituting the values yields

$$k_v = \frac{1}{2}\dot{q}^T\begin{bmatrix} k_{v11} & k_{v12} \\ k_{v21} & k_{v22} \end{bmatrix}\dot{q}, \tag{7.45}$$

where

$$\begin{aligned} k_{v11} = {} & M(l_1^2 + l_2^2) + 2l_1(\cos(q_1)\cos(q_1+q_1) + \sin(q_1)\sin(q_1+q_1)) \\ & (Ml_2 + m_2 l_{c2}) + m_2(l_1^2 + l_{c2}^2) + m_1 l_{c1}^2, \end{aligned} \tag{7.46}$$

$$\begin{aligned} k_{v12} = k_{v21} = {} & Ml_2^2 + m_2 l_{c2}^2 + l_1(\cos(q_1)\cos(q_1+q_1) \\ & + \sin(q_1)\sin(q_1+q_1))(Ml_2 + m_2 l_{c2}), \end{aligned} \tag{7.47}$$

$$k_{v22} = Ml_2 + m_2 l_{c2}^2. \tag{7.48}$$

The rotational part of the kinetic energy is:

$$k_w = \frac{1}{2}\dot{q}^T\left\{J_{wc1}^T \;\; R_1^0 \; I_1 \;\; R_1^{0T} \;\; J_{wc1} + J_{wc2}^T \;\; R_2^0 \; I_2 \; R_2^{0T} \;\; J_{wc2} + J_{w2}^T \;\; R_2^0 \; I_m \; R_2^{0T} \;\; J_{w2}\right\}\dot{q}, \tag{7.49}$$

$$\begin{aligned} J_{wc1}^T \;\; R_{c1} \; I_1 \; R_{c1}^T \; J_{wc1} = I_1 & \begin{bmatrix} 0 & 0 & 1 \\ 0 & 0 & 0 \end{bmatrix}\begin{bmatrix} \cos(q_1) & -\sin(q_1) & 0 \\ \sin(q_1) & \cos(q_1) & 0 \\ 0 & 0 & 1 \end{bmatrix} \\ & \begin{bmatrix} \cos(q_1) & \sin(q_1) & 0 \\ -\sin(q_1) & \cos(q_1) & 0 \\ 0 & 0 & 1 \end{bmatrix}\begin{bmatrix} 0 & 0 \\ 0 & 0 \\ 1 & 0 \end{bmatrix} = I_1\begin{bmatrix} 1 & 0 \\ 0 & 0 \end{bmatrix}, \end{aligned} \tag{7.50}$$

$$\begin{aligned} J_{wc2}^T \;\; R_{c2} \; I_2 \; R_{c2}^T \; J_{wc2} = I_2 & \begin{bmatrix} 0 & 0 & 1 \\ 0 & 0 & 1 \end{bmatrix}\begin{bmatrix} \cos(q_1) & -\sin(q_1) & 0 \\ \sin(q_1) & \cos(q_1) & 0 \\ 0 & 0 & 1 \end{bmatrix} \\ & \begin{bmatrix} \cos(q_1) & \sin(q_1) & 0 \\ -\sin(q_1) & \cos(q_1) & 0 \\ 0 & 0 & 1 \end{bmatrix}\begin{bmatrix} 0 & 0 \\ 0 & 0 \\ 1 & 1 \end{bmatrix} = I_2\begin{bmatrix} 1 & 1 \\ 1 & 1 \end{bmatrix}, \end{aligned} \tag{7.51}$$

$$J_{w2}^T \;\; R_2 \; I_m \; R_2^T \; J_{w2} = I_m\begin{bmatrix} 0 & 0 & 1 \\ 0 & 0 & 1 \end{bmatrix}\begin{bmatrix} \cos(q_1) & -\sin(q_1) & 0 \\ \sin(q_1) & \cos(q_1) & 0 \\ 0 & 0 & 1 \end{bmatrix}$$

$$\begin{bmatrix} \cos(q_1) & \sin(q_1) & 0 \\ -\sin(q_1) & \cos(q_1) & 0 \\ 0 & 0 & 1 \end{bmatrix} \begin{bmatrix} 0 & 0 \\ 0 & 0 \\ 1 & 1 \end{bmatrix} = I_m \begin{bmatrix} 1 & 1 \\ 1 & 1 \end{bmatrix}. \tag{7.52}$$

By substituting (7.50), (7.51), and (7.52) into (7.49), we get the overall rotational part of the kinetic energy, which can be written as follows:

$$k_w = \frac{1}{2}\dot{q}^T \left\{ \begin{bmatrix} I_1 + I_2 + I_m & I_2 + I_m \\ I_2 + I_m & I_2 + I_m \end{bmatrix} \right\} \dot{q}. \tag{7.53}$$

The overall kinetic energy can be obtained by adding (7.45)–(7.53), and it will be written in the following matrix form:

$$K = \frac{1}{2}\dot{q}^T \{D(q)\}\dot{q}, \tag{7.54}$$

where $D(q)$ is a 2×2 matrix and takes the form

$$D(q) = \begin{bmatrix} d_{11} & d_{12} \\ d_{21} & d_{22} \end{bmatrix}, \tag{7.55}$$

where

$$\begin{aligned} d_{11} = {} & M(l_1^2 + l_2^2) + 2l_1(\cos(q_1)\cos(q_1 + q_1) + \sin(q_1)\sin(q_1 + q_1))(Ml_2 + m_2 l_{c2}) \\ & + m_2(l_1^2 + l_{c2}^2) + m_1 l_{c1}^2 + I_1 + I_2 + I_m, \end{aligned} \tag{7.56}$$

$$\begin{aligned} d_{12} = d_{21} = {} & Ml_2^2 + m_2 l_{c2}^2 + l_1(\cos(q_1)\cos(q_1 + q_1) \\ & + \sin(q_1)\sin(q_1 + q_1))(Ml_2 + m_2 l_{c2}) + I_2 + I_m, \end{aligned} \tag{7.57}$$

$$d_{22} = Ml_2 + m_2 l_{c2}^2 + I_2 + I_m. \tag{7.58}$$

The potential energy is as follows:

$$P_1 = m_1 g l_{c1} \sin(q_1), \tag{7.59}$$

$$P_2 = m_2 g(l_1 \sin(q_1) + l_{c2}\sin(q_1 + q_2)), \tag{7.60}$$

$$P_3 = Mg(l_1 \sin(q_1) + l_2 \sin(q_1 + q_2)). \tag{7.61}$$

The total potential energy will be the summation of (7.59), (7.60), and (7.61):

$$P = m_1 g l_{c1} \sin(q_1) + m_2 g(l_1 \sin(q_1) + l_{c2}\sin(q_1 + q_2)) + Mg(l_1 \sin(q_1)$$

$$+ l_2 \sin(q_1 + q_2) \tag{7.62}$$

Now, we can estimate the Lagrangian of the robot arm L, which is the difference between kinetic and potential energies:

$$L = K - P. \tag{7.63}$$

Now, we are ready to drive the dynamic equations for the robot using Eq. (7.9)

$$a_1\ddot{q}_1 + b_1\ddot{q}_2 + h_1\dot{q}_1\dot{q}_2 + k_1\dot{q}_2^2 + p_1 = \tau_1 \tag{7.64}$$

$$a_2\ddot{q}_1 + b_2\ddot{q}_2 + k_2\dot{q}_1^2 + p_2 = \tau_2, \tag{7.65}$$

where

$$a_1 = m_1 l_{c1}^2 + m_2\left(l_1^2 + l_{c2}^2 + 2l_1 l_{c2}\cos(q_2)\right) + M\left(l_1^2 + l_2^2 + 2l_1 l_2 \cos(q_2)\right)$$
$$+ I_1 + I_2 + I_m, \tag{7.66}$$

$$b_1 = a_2 = m_2\left(l_{c2}^2 + l_1 l_{c2}\cos(q_2)\right) + M\left(l_2^2 + l_1 l_2 \cos(q_2)\right) + I_2 + I_m, \tag{7.67}$$

$$h_1 = -2l_1 \sin(q_2)(l_{c2}m_2 + l_2 M), \tag{7.68}$$

$$k_1 = 2h_1, \tag{7.69}$$

$$p_1 = g(l_{c1}m_1 + l_1 m_2 + l_1 M)\cos(q_1) + g(l_{c2}m_2 + l_2 M)\cos(q_1 + q_2), \tag{7.70}$$

$$b_2 = l_{c2}^2 m_2 + l_2^2 M + I_2 + I_m, \tag{7.71}$$

$$k_2 = (m_2 l_{c2} + M l_2) l_1 \sin(q_2), \tag{7.72}$$

$$p_2 = g(m_2 l_{c2} + M l_2)\cos(q_1 + q_2), \tag{7.73}$$

MATLAB code

```
clear;
clc;
syms th1(t) th2(t) L1 L2 Lc1 Lc2 m1 m2  M w1(t) w2(t) g
syms I1xx I1xy I1xz I1yx I1yy I1yz I1zx I1zy I1zz
I1=[I1xx I1xy I1xz;I1yx I1yy I1yz;I1zx I1zy I1zz];
syms I2xx I2xy I2xz I2yx I2yy I2yz I2zx I2zy I2zz
I2=[I2xx I2xy I2xz;I2yx I2yy I2yz;I2zx I2zy I2zz];
syms Imxx Imxy Imxz Imyx Imyy Imyz Imzx Imzy Imzz
Im=[Imxx Imxy Imxz;Imyx Imyy Imyz;Imzx Imzy Imzz];

alpha1 = 0;
alpha2 = 0;
d1 = 0;
d2 = 0;
tc1 = Denavit(th1,alpha1,Lc1,d1); % HTM at the center of
mass of link 1
tc1mat = tc1(t);
pc1 = tc1mat(1:3,4); % position vector of the center of
mass of link 1
Jvc1
=[functionalDerivative(pc1(1),th1),functionalDerivative(pc
1(1),th2);...

functionalDerivative(pc1(2),th1),functionalDerivative(pc1(
2),th2);...

functionalDerivative(pc1(3),th1),functionalDerivative(pc1(
3),th2)]; % Jacobian at the center of mass of link 1
z1 = tc1mat(1:3,3); % Z-direction at the center of mass of
link 1
Jwc1 = [z1,0];
R1 = tc1mat(1:3,1:3); % Rotational matrix at the center of
mass of link 1
%%%%%%%%%%%%%%%%%%%%%%%%
t1 = Denavit(th1,alpha1,L1,d1);
t1mat = t1(t);
tc2 = Denavit(th2,alpha2,Lc2,d2);
tc2mat = tc2(t);
t_ofc2 = t1mat*tc2mat;% HTM at the center of mass of link
2
t_ofc2mat = t_ofc2;
pc2 = t_ofc2mat(1:3,4);% position vector of the center of
mass of link 2
R2 = t_ofc2mat(1:3,1:3);% Rotational matrix of the center
of mass of link 2
Jvc2
=[functionalDerivative(pc2(1),th1),functionalDerivative(pc
```

```
2(1),th2);...

functionalDerivative(pc2(2),th1),functionalDerivative(pc2(
2),th2);...

functionalDerivative(pc2(3),th1),functionalDerivative(pc2(
3),th2)];
z2 = tc2mat(1:3,3); % Z-direction at the center of mass of
link 1
Jwc2 = [z1,z2];
%%%%%%%%%%%%%%%%%%%%%%
t2 = Denavit(th2,alpha2,L2,d2);
t2mat = t2(t);
t_tip = t1mat*t2mat;% HTM at the tip of link 2
% t_tipmat = t_tip(t);
p2 = t_tip(1:3,4);% position vector of the m3
Jv2
=[functionalDerivative(p2(1),th1),functionalDerivative(p2(
1),th2);...

functionalDerivative(p2(2),th1),functionalDerivative(p2(2)
,th2);...

functionalDerivative(p2(3),th1),functionalDerivative(p2(3)
,th2)];
Jw2 = t2mat(1:3,3); % Z-direction at the center of mass of
link 1
P = m1*g*Lc1*sin(th1) + m2*g*L1*sin(th1)+
m2*g*Lc2*sin(th1+th2)+...
    M*g*L1*sin(th1)+ M*g*L2*sin(th1+th2);
w = [w1;w2];
tran = m1*Jvc1.'*Jvc1+m2*Jvc2.'*Jvc2+M*Jv2.'*Jv2;
kw1 = Jwc1.'*R1*I1*R1.'*Jwc1;
kw2 = Jwc2.'*R2*I2*R2.'*Jwc2;
kwm = Jwc2.'*R2*Im*R2.'*Jwc2;
rota = kw1+kw2+kwm;
kin = 0.5*w.'*(tran+rota)*w;
lagran = kin-P;
torque_1 = diff(functionalDerivative(lagran,w1))-
functionalDerivative(lagran,th1);
torque_2 = diff(functionalDerivative(lagran,w2))-
functionalDerivative(lagran,th2)
```

The function Denavit:

```
function [DH] = Denavit(theta,alpha,a,d)
DH = [cos(theta),-
1*sin(theta)*cos(alpha),sin(theta)*sin(alpha),a*cos(theta)
;...
    sin(theta),cos(theta)*cos(alpha),-
1*cos(theta)*sin(alpha),a*sin(theta);...
    0,sin(alpha),cos(alpha),d;...
    0,0,0,1];
end
```

Note that *sin* and *cos* functions are specified for radian measurements and symbolic functions and differ from *sind* and *cosd*, which are used for degree unit. The inertia tensor is given in the centre of mass of the link.

7.5 Inertia Tensor

The inertia matrix of a body is constant regardless of the motion and is usually calculated to a frame attached to that body. In Example 7.1, we have transformed the representation of the inertia tensor to the inertial frame using a rotational matrix by similarity transformation in Eq. (7.25) and that makes the inertia tensor depend on the configuration of the robot. Generally, the inertia tensor is a 3×3 matrix taking the form

$$I = \begin{bmatrix} I_{xx} & I_{xy} & I_{xz} \\ I_{yx} & I_{yy} & I_{yz} \\ I_{zx} & I_{zy} & I_{zz} \end{bmatrix}, \tag{7.74}$$

where

$$I_{xx} = \iiint (y^2 + z^2)\rho(x, y, z)\,\mathrm{d}x\,\mathrm{d}y\,\mathrm{d}z$$

$$I_{yy} = \iiint (x^2 + z^2)\rho(x, y, z)\mathrm{d}x\,\mathrm{d}y\,\mathrm{d}z$$

$$I_{zz} = \iiint (x^2 + y^2)\rho(x, y, z)\mathrm{d}x\,\mathrm{d}y\,\mathrm{d}z$$

$$I_{xy} = I_{yx} = -\iiint xy\,\rho(x, y, z)\mathrm{d}x\,\mathrm{d}y\,\mathrm{d}z$$

$$I_{xz} = I_{zx} = -\iiint xz\,\rho(x, y, z)\mathrm{d}x\,\mathrm{d}y\,\mathrm{d}z$$

$$I_{yz} = I_{zy} = -\iiint yz\,\rho(x, y, z)\mathrm{d}x\,\mathrm{d}y\,\mathrm{d}z$$

and $\rho(x, y, z)$ is mass density as a function of distance.

The initial and final values of the integrals above are the boundaries of the space occupied by the body. Principal moments of inertia I_{xx}, I_{yy}, and I_{zz} calculated about the x-, y-, and z-axes, respectively. Off-diagonal elements of the inertia matrix are called cross-products of inertia, where they take the value of zero in case the rigid body is symmetric about an axis attached to that body.

7.6 Discussion

It is very essential to understand the logic behind the dynamic equations described in this chapter, especially when we would like to design the control system of a robot. First of all, Euler–Lagrange formulation is not the only method to describe dynamic equations, there also exists Newton–Euler formulation, and both of the two methods lead to the same results such as Eqs. (7.64) and (7.65). Euler–Lagrange formulation treats the robot as a whole; in reverse, Newton–Euler formulation treats the robot as separated links. Anyway, in this book we do not see a reason to describe Newton–Euler method since the reader can find it in other books. The dynamic equations driven by these two methods are ideal, the equations describing the dynamic behaviour of a chain of rigid bodies. In real life, we cannot see rigid links and components, and there is always elastic deformation in links and bearings. In Chap. 8, we introduced this problem and presented methods to find the elastic deformation in the components of the robot. The elastic deflections have a sensible effect on the potential energy which we have dropped them from our calculations in this chapter. Also, friction forces in joints affect the value of kinetic energy of the system; in turn, friction forces should be considered when one thinks to design a control system. In general, an extensive dynamic analysis including elasticity, friction, vibrations, damping, etc., gives real model far from the idealized one.

7.7 Summary

In this chapter, we have presented Euler–Lagrange formulation to find the dynamic equations for the robot manipulator. This ideal formulation is based on the assumption that we deal with rigid bodies and bearing. To give the reader significant insight, we have presented a problem, and all its parameters are calculated from the previous calculation which are path planning, trajectory planning, Jacobian, and inverse kinematics. Also, we have introduced MATLAB symbolic toolbox to drive the dynamic equations of the robot. This toolbox has significant benefits, especially when the designer would like to drive dynamic equations in a systematic and automatic way.

Exercises

7.1 For the 3DOF, two revolute joints and one prismatic joint RRP manipulator in Fig. 7.5, find the dynamic equations at the moment that the robot is at point (136,100). Assume any missing information that you may need.

7.2 For the robot in Problem 7.1, find the force history of the actuators assuming that the robot moves from (197,15) to (27,198) in 2 s by writing an appropriate program. Assume any missing information that you may need.

7.3 Assume that three DC motors are actuating the robot in Problem 7.1 and each one has a maximum speed of 500 rpm. What is the appropriate time to pass the trajectory from the start point (197,15) to the goal point (27,198), and is it more than 2 s or less than 2 s.

7.4 Consider the density of the material of the links is 8050 kg/m^3, and what is the limitation on the forces of the actuators.

7.5 For the manipulator in Fig. 7.6, discuss the problems mentioned in 7.2.

7.3 Repeat Example 7.1 by assuming that there is a total elastic deformation of 10 mm.

7.4 Repeat Example 7.1 by assuming that there are total friction forces at the bearings $F = 0.2\,\dot{q}$, where $\dot{q}$ is the angular velocity of the joint.

7.5 Drive the dynamic equations of the robot in Example 4.2.

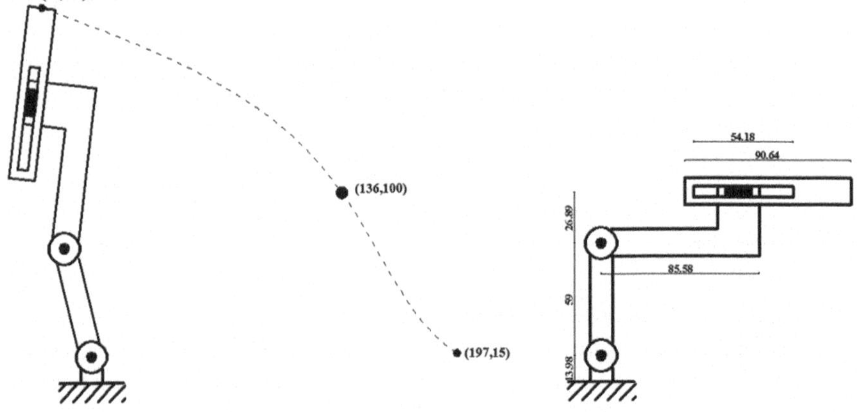

Fig. 7.5 RRP manipulator

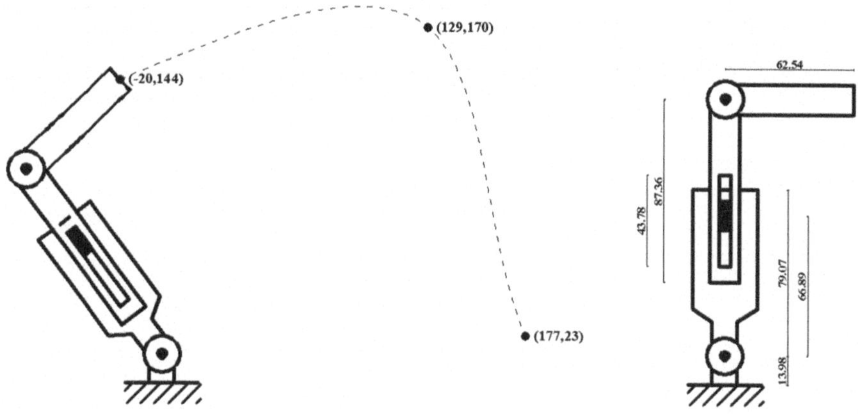

Fig. 7.6 RPR manipulator

Reference

Ghafil H, Jármai K (2019) Optimum dynamic analysis of a robot arm using flower pollination algorithm. In: Advances and trends in engineering sciences and technologies III: proceedings of the 3rd international conference on engineering sciences and technologies (ESaT 2018). High Tatras Mountains, Tatranské Matliare, 12–14 Sep 2018, Slovak Republic, CRC Press, p 81

Chapter 8
Structural Optimization and Stiffness Analysis

8.1 Introduction

One of the most critical issues in robotics is the optimization of the dimensions of the physical components of the robot and thus mass reduction. This issue can be dealt with by a kinematic and kinetic viewpoint. In this chapter, we will separate the problem into two subproblems. The first one is finding optimum link lengths, and the second is finding the optimum thicknesses of the links. In this chapter, the links are assumed to have uniform cross-sections. The first part of the chapter focuses on finding optimum lengths using kinematic equations, while the second part is about using stiffness analysis to calculate the compliant displacements or deformations due to applying external forces. This is very important, especially for accurate position calculations, vibration control, fatigue analysis, and mechanical design of the arm.

8.2 Structural Optimization

In the kinematic synthesis of a multilinkage robot arm, topology optimization is performed to obtain the best type of synthesis for a single-task or multitask robot manipulator. Next, the minimum required dimensions should be calculated during the dimensional synthesis process. Both the type and dimensional synthesis can be done in two separate operations or in one single operation. Topology optimization means finding the minimum required links and joints of a robot for a specific function or multitask regardless of the dimensions. Once we have the proper topology (robot type), we have to search for the best dimensions of the linkages of the robot—this operation is called dimensional synthesis. In this section, we assume that we have the best robot type and focus only on finding optimum dimensionality. Specifically, in this section we will explain how to find optimum lengths of the links of the robot (Ghafil and Jármai 2019). Reduction of the lengths of the links leads to a reduction in the required masses and weights, and this dramatically affects the speed of the

H. N. Ghafil and K. Jármai (ed.), *Optimization for Robot Modelling with MATLAB*,
https://doi.org/10.1007/978-3-030-40410-9_8

robot as well as decreasing the manufacturing costs of the manipulator. Assuming that the links have hollow cross-sections, our target is to find the minimum required length and wall thickness.

8.2.1 Structural Optimization of a 7DOF-Type Robot

Recall the 7DOF manipulator in Example 4.4. We can follow the same procedure as in Example 4.6 to find the inverse kinematic position. For a specific Cartesian point, we have to obtain the proper set of joint variables given spatial parameters of the links, i.e. α, a, and d. We can modify the problem to the optimization algorithm to search for the proper set of joint variables as well as the proper a and d parameters. In other words, for a given Cartesian point, each link is represented by three variables, θ, a, and d where a and d are proposed link length and link offset. Angle of twist is related to the robot type, and that is why we do not employ it in our calculations. In other words, parameters a and d are associated with the dimensional synthesis (structural optimization), while α is related to the type synthesis (topology optimization). Back to Example 4.4, we have assumed that $a = 600$, $d_1 = 1000$, and $d_7 = 400$ and they are all that we need to describe the dimensionality of the manipulator. For the purpose of the structural optimization, we assume that these parameters are unknown as well as the set of the seven joint variables. In total, we have ten variables that should be optimized, where the lower and upper limits for each variable are shown as follows:

$$
\begin{aligned}
&-160 \le \theta_1 \le 160, \quad -150 \le \theta_2 \le 30, \quad -120 \le \theta_3 \le 120, \quad -45 \le \theta_4 \le 130,\\
&-30 \le \theta_5 \le 180, \quad -90 \le \theta_6 \le 90, \quad -100 \le \theta_7 \le 100,\\
&500\ \text{mm} \le d_1 \le 1000\ \text{mm}, \quad 300\ \text{mm} \le a \le 700\ \text{mm}, \quad 200\ \text{mm} \le d_7 \le 400\ \text{mm}.
\end{aligned}
$$

Assume that we need to find the best dimensions that enable the robot to reach the task point (1560, 0, 1005) in a space confined by 2500 in x-direction, 1000 in y-direction, and 1500 in z-direction. The artificial bee colony in Example 4.6 can be modified to solve this issue as follows:

```
nVar=10;
VarSize=[1 nVar];
VarMin= [-160,-150,-120,-45,-30,-90,-100,500,300,200];
% Decision Variables Lower Bound
VarMax= [160,30,120,130,180,90,100,1000,700,400];
% Decision Variables Upper Bound
```

The cost function that should be minimized, `tunaobj`, takes ten inputs and returns one output, which is the cost of the proposed solution.

```
function z = tunaobj(sol)
t1 = sol(1);
t2 = sol(2);
t3 = sol(3);
t4 = sol(4);
t5 = sol(5);
t6 = sol(6);
t7 = sol(7);
d1 = sol(8);
a  = sol(9);
d7 = sol(10);
[x,y,z,J] = tun_for(t1,t2,t3,t4,t5,t6,t7,d1,a,d7);
z = sqrt((x-1560)^2+(y)^2+(z-1005)^2);
         if(abs(J.x)>2500 || abs(J.y)>1000||
abs(J.z)>1500)
           z=z+1000;
         end
end
```

`tun_for` is the function that generates the corresponding overall homogeneous transformation matrix and returns the position vector of the matrix. In addition, this function returns the position vector of the joint next to the end-effector `J` for the purpose of collision detection with the limited working space.

```
function [x,y,z,J] =
tun_for(t1,t2,t3,t4,t5,t6,t7,d1,a,d7)
format short G
th_1= t1;      % joint angle 1
tw_1 = -90;        %twist angle for link 1
a_1 = 0;           %distance between z1 and z2
d_1 = d1;        % distance between x1 and x2
th_2= t2;      % joint angle 2
tw_2 = 90;        %twist angle for link 2
a_2 = a;        %distance between z2 and z3
d_2 = 0;          % distance between x2 and x3
th_3= 90+t3;      % joint angle 3
tw_3 = 0;         %twist angle for link 3
a_3 = a;        %distance between z3 and z4
d_3 = 0;        % distance between x3 and x4
th_4= t4-90;      % joint angle 4
tw_4 = 0;         %twist angle for link 4
a_4 = a;        %distance between z4 and z5
d_4 = 0;          % distance between x4 and x5
th_5= t5-90;      % joint angle 5
```

```
tw_5 = -90;         %twist angle for link 5
a_5 = a;           %distance between z5 and z6
d_5 = 0;             % distance between x5 and x6
th_6= t6;        % joint angle 6
tw_6 = -90;          %twist angle for link 6
a_6 = 0;             %distance between z6 and z7
d_6 = 0;             % distance between x6 and x7
th_7= t7;        % joint angle 7
tw_7 = 0;            %twist angle for link 7
a_7 = 0;             %distance between z7 and zn
d_7 = d7;           % distance between x7 and xa
H_1= Denavit(th_1,tw_1,a_1,d_1);
H_2= Denavit(th_2,tw_2,a_2,d_2);
H_3= Denavit(th_3,tw_3,a_3,d_3);
H_4= Denavit(th_4,tw_4,a_4,d_4);
H_5= Denavit(th_5,tw_5,a_5,d_5);
H_6= Denavit(th_6,tw_6,a_6,d_6);
H_7= Denavit(th_7,tw_7,a_7,d_7);
TH = H_1*H_2*H_3*H_4*H_5*H_6*H_7;
x=TH(1,4);
y=TH(2,4);
z=TH(3,4);
C = H_1*H_2*H_3*H_4*H_5*H_6;
J.x = C(1,4);
J.y = C(2,4);
J.z = C(3,4);
end
```

The optimal solution for this problem $\left[\theta_1\ \theta_2\ \theta_3\ \theta_4\ \theta_5\ \theta_6\ \theta_7\ d_1\ a\ d_7\right]$ implicitly includes the optimal lengths of the links as well as the optimum set of joint angles. Also, this solution is suitable for a single-task manipulator and is called kinematic-based structural optimization because it is based on the kinematic equations. Thus, a structure optimized by this method represents a geometric solution and does not reflect any effects of internal or external forces. Furthermore, to find the best dimensionality of a robot that satisfies a given trajectory, we have to enumerate all of the corresponding dimensionalities for all the given points on the trajectory. For example, the trajectory in Fig. 8.1 has five Cartesian points, and a robot (with proper dimensions) has to access all of these points. For this issue, we have to find a solution (robot dimensions) for each point and then choose the one that can satisfy all given points of the trajectory or trajectories.

Figure 8.2 expresses five possible configurations, each corresponding to a single point on the trajectory. Optimization algorithms return their solution for the best dimensional synthesis of the robot type that satisfies all points on the given trajectory.

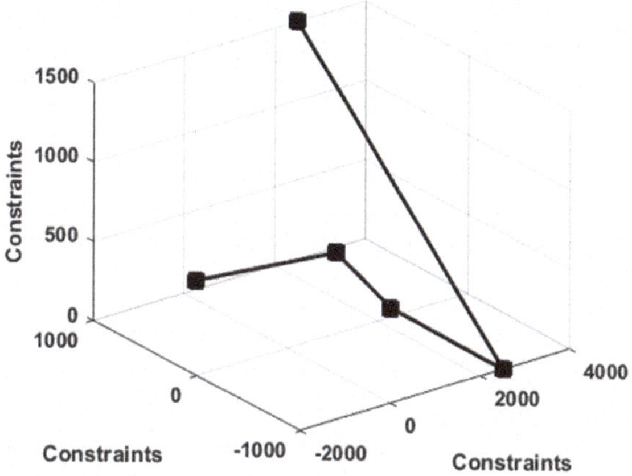

Fig. 8.1 Sample of a trajectory

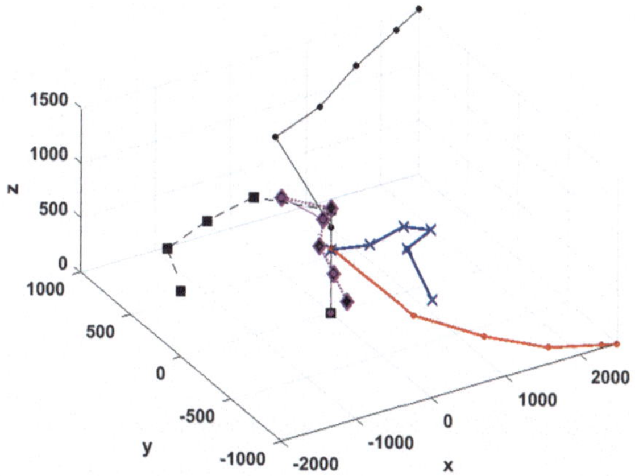

Fig. 8.2 Corresponding configurations with different dimensions for the trajectory in Fig. 8.1

8.3 Stiffness Analysis

Stiffness analysis (Gonçalves et al. 2016) of a manipulator is an open issue in robotics which has been widely discussed in the literature, with different methods being proposed. Stiffness of a mechanical system is its capacity to undergo external loading and show elastic deformation to that loading. Only elastic deformation or compliant displacement is considered in this analysis; i.e. stiffness calculation is not valid for a mechanical system that experiences plastic compliant displacements. From the viewpoint of robotics, compliant displacements have a negative effect on the performance of the robot because they cause problems in the accurate positioning of the end-effector (vibration). Other side effects of compliant displacements are fatigue strength, static behaviour, and friction losses (Gonçalves et al. 2016). Stiffness analysis is essential for the design of manipulators since it enables us to choose materials and proper kinematic synthesis properly. In other words, the study of stiffness enables us to calculate the optimum wall thickness in case we have a hollow cross-sectional area of the link.

The overall stiffness matrix depends on the following parameters (Tsai 1999):

1. Young's modulus of elasticity of the material,
2. Dimensions of the mechanical system,
3. Transmission mechanisms,
4. Configuration of the robot (Yoon et al. 2004).

Using large and heavy parts leads to increasing the stiffness performance and decreasing the motion of the system; a balance between these two issues should be considered. In this section, two methodologies will be described to calculate the stiffness matrix of a manipulator. When wrench ΔF acts upon an end-effector of a robotic system in the equilibrium state, a compliant displacement Δx is imposed in the form of rotational and translational displacements.

$$\Delta F = k\,\Delta x, \tag{8.1}$$

where the wrench is a 1×6 vector $\Delta F = \begin{bmatrix} F_x \\ F_y \\ F_z \\ M_x \\ M_y \\ M_z \end{bmatrix}$, k is a 6×6 stiffness matrix, and compliant displacement is represented by a 1×6 vector $\Delta x = \begin{bmatrix} \delta_x \\ \delta y \\ \delta_z \\ \phi_x \\ \phi_y \\ \phi_z \end{bmatrix}$.

Equation (8.1) is a direct application of Hook's law for small deformations where they are proportional to the applied forces. It is worth mentioning that the stiffness matrix does not have any meaning by itself, and stiffness is a physical property of material under loading conditions. The stiffness analysis of a robot provides us with the stiffness matrix for each configuration. In other words, the deformation of the robotic arm is different from one configuration to another for fixed mounts of applied wrenches. Consequently, the robot controller should consider, for example, reducing the motion speed for configurations that drive the robot to have large values of the corresponding compliant displacement. Figure 8.3 shows the possible force and moments acting on the terminal ends of a link, while Fig. 8.4 illustrates the compliant displacements corresponding to the acting wrench. For given amounts of wrench, we can calculate the corresponding deformations depending on Eq. (8.1).

$$\Delta x = k^{-1} \Delta F = C\,\Delta F, \tag{8.2}$$

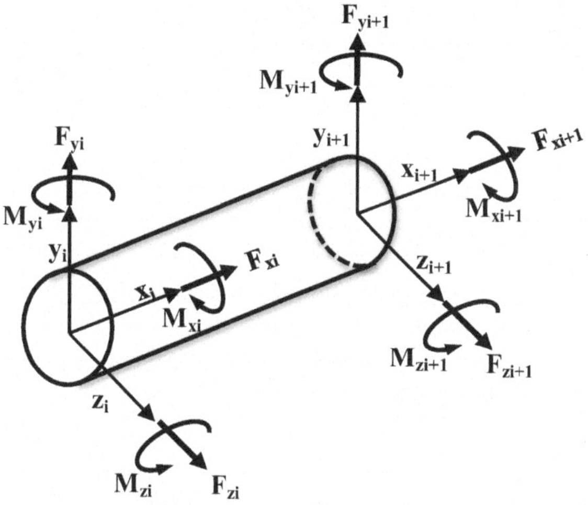

Fig. 8.3 Forces and moments acting on a link

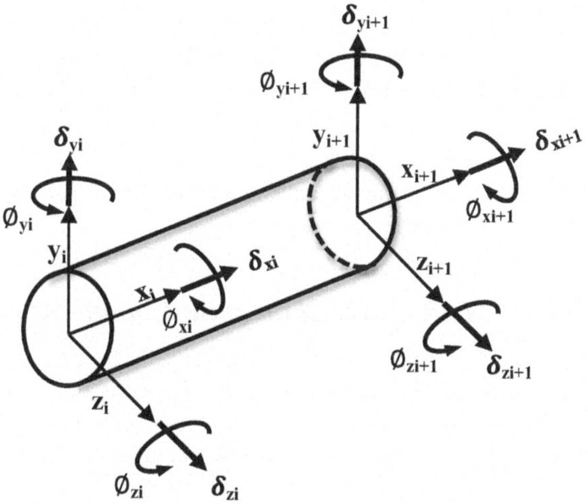

Fig. 8.4 Translational and rotational displacements due to forces and moments acting on a link

where C is the inverse of the stiffness matrix and called the compliance matrix and sometimes also the flexibility matrix. The joint, actuators, and links are the main sources of this matrix. Some methods consider only the stiffness of the joints or the links or even both of them, and for simplicity some of the links can be considered as rigid links. From the theory of elasticity, we can find the 6×6 stiffness matrix

$$k = \begin{bmatrix} \frac{A_j E_j}{\ell_j} & 0 & 0 & 0 & 0 & 0 \\ 0 & \frac{12E_j I_{zj}}{\ell^3{}_j} & 0 & 0 & 0 & \frac{6E_j I_{zj}}{\ell^2{}_j} \\ 0 & 0 & \frac{6E_j I_{yj}}{\ell^3{}_j} & 0 & -\frac{6E_j I_{yj}}{\ell^2{}_j} & 0 \\ 0 & 0 & 0 & \frac{G_j J_j}{\ell_j} & 0 & 0 \\ 0 & 0 & -\frac{6E_j I_{yj}}{\ell^2{}_j} & 0 & \frac{4E_j I_{yj}}{\ell_j} & 0 \\ 0 & \frac{6E_j I_{zj}}{\ell^2{}_j} & 0 & 0 & 0 & \frac{4E_j I_{zj}}{\ell_j} \end{bmatrix}, \tag{8.3}$$

where A is the cross-sectional area of the link j, E is the modulus of elasticity, G is the shear modulus, and ℓ is the length of the link.

I_{yj} and I_{zj} are moments of areas about y- and z-axes, respectively. J_j is the Saint-Venant torsion constant.

8.4 Lumped Parameter Model

The lumped parameter model or lumped element model assumes that the joints are the main source of the compliant displacements. The links are assumed to be rigid with concentrated mass at the centre, and they interact with each other via links. Lumped parameter methods are divided into two groups: Methods depend on kinematic and dynamic forces and methods which use Jacobians. In this section, two types of formulations are introduced using the Jacobian matrix (Komatsu 1990).

8.4.1 Method Based on Joint Compliance

This method takes into account the joints as the only source of the compliant displacements regardless of other sources, as described by Tsai (1999). From Hook's law for small angular deformation, we have

$$\tau_i = k_i \, \Delta q_i, \tag{8.4}$$

where τ_i is the applied torque to the joint, k_i is the stiffness component of the joint, and Δq_i is the angular displacement achieved by the joint. For a mechanical system, we may write that in the matrix form

$$\tau = K_d \, \Delta q, \tag{8.5}$$

where $\tau = \left[\, \tau_1 \; \tau_2 \cdots \tau_n \,\right]$, $K_d = \text{diag}(k_{d1}, k_{d2}, \ldots, k_{dn})$, and $\Delta q = \left[\, \Delta q_1 \; \Delta q_2 \cdots \Delta q_n \,\right]$.

From study of the robot kinematics, we map the forces and displacements from Cartesian space to the joint space by the Jacobian

$$\Delta x_{jt} = J_{jt} \, \Delta q \tag{8.6}$$

$$\tau = J_{jt}^T \, F, \tag{8.7}$$

where Δx_{jt} is compliant displacements experienced by the end-effector due only to compliant joints. J_{jt} is the Jacobian defined by $J_{jt} = \frac{\partial f}{\partial q_i}$, where f is the position of the end-effector and q_i is joint variables. The subscript *jt* for Jacobian is used to be consistent with the next section. From Eqs. (8.5), (8.6), and (8.7), we get the expression shown in Eq. (8.2) that relates the external wrenches ΔF to the compliant displacements Δx_{jt}

$$\Delta x_{jt} = C_{jt} \, \Delta F \tag{8.8}$$

where C_{jt} is the compliance matrix of the system having only compliant joints

$$C_{jt} = J_{jt} K_d^{-1} J_{jt}^T \tag{8.9}$$

The global stiffness matrix for the whole manipulator is the inverse of this:

$$K_s = C_{jt}^{-1} \tag{8.10}$$

8.4.2 Method Based on Joints and Links Compliants

The compliant link was included in the previous calculations as well as the compliant joints by Komatsu (1990). This formulation states that the global compliance matrix C_T is the summation of the compliance matrix obtained by compliant joints C_{jt} and the compliance matrix obtained due to compliant links C_ℓ. In Sect. 8.4.1, we have learned how to find C_{jt} while C_ℓ is defined by

$$C_\ell = J_\ell K_\ell^{-1} J_\ell^T, \tag{8.11}$$

where $K_\ell = \mathrm{diag}(k_{\ell 1}, k_{\ell 2}, \ldots, k_{\ell n}), k_{\ell i} (i = 1, \ldots, n)$ is defined by Eq. (8.3) and the Jacobian of the elastic deformation J_ℓ is obtained by

$$J_\ell = \frac{\partial f}{\partial \delta} \tag{8.12}$$

where $\delta = \left(\delta_1 \; \delta_2 \; \cdots \; \delta_n\right)^T$ is the translational displacements of the origins of the attached frames to the manipulator. Thus, the total compliance matrix is

$$C_T = C_\ell + C_{jt} \tag{8.13}$$

The global stiffness matrix can be obtained using Eq. (8.10).

Example 8.1 Consider the 2DOF planar manipulator in Fig. 8.5, where we need to find the stiffness matrix using the Komatsu formulation for the given external wrenches F_x and F_y. Figure 8.6a, b shows the linear and angular displacements to the joint 1 (point A) and end-effector (point B), respectively.

Firstly, we have to derive the coordinates of the end-effector position:

$$x_A = \ell_1 \cos(\theta_1) - \delta_1 \sin(\theta_1) \tag{8.14}$$

$$y_A = \ell_1 \sin(\theta_1) + \delta_1 \cos(\theta_1) \tag{8.15}$$

$$x_B = x_A + \ell_2 \cos(\theta_1 + \phi_1 + \theta_2) - \delta_2 \sin(\theta_1 + \phi_1 + \theta_2) \tag{8.16}$$

$$y_B = y_A + \ell_2 \sin(\theta_1 + \phi_1 + \theta_2) - \delta_2 \cos(\theta_1 + \phi_1 + \theta_2) \tag{8.17}$$

The perpendicular forces acting on link 1 and link 2 are F_1 and F_2, respectively:

$$F_1 = F_x \sin(\theta_2) + F_y \cos(\theta_2) \tag{8.18}$$

$$F_2 = F_y \tag{8.19}$$

Moments about point B are M_1 and M_2, respectively:

$$M_1 = \ell_2 F_y \tag{8.20}$$

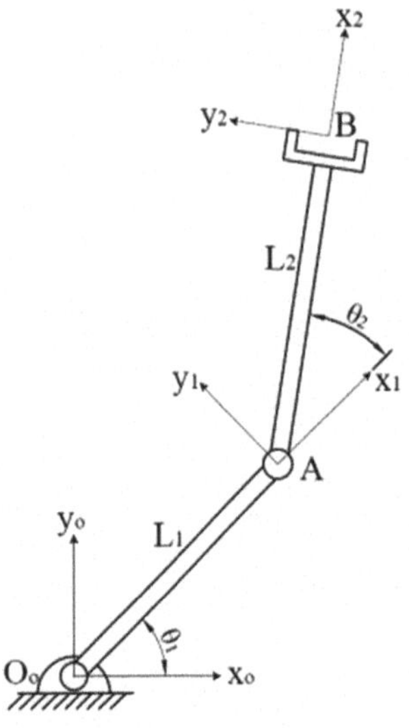

Fig. 8.5 2DOF manipulator

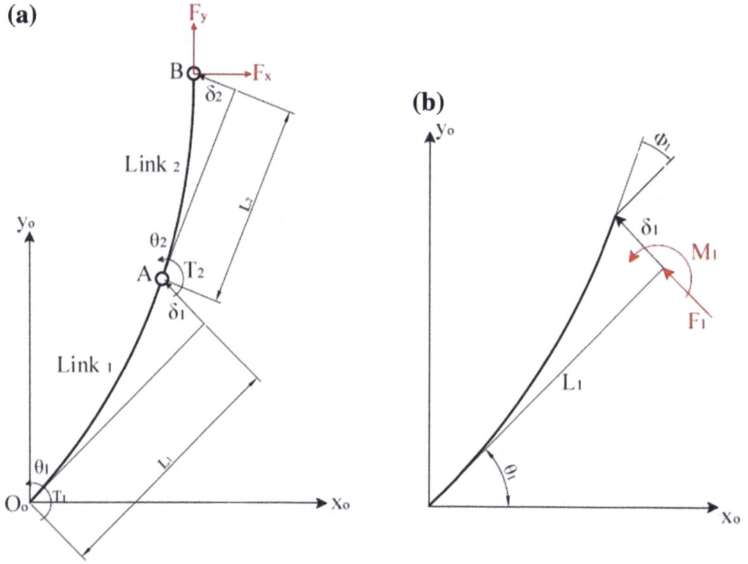

Fig. 8.6 **a** Translational compliant displacements δ and **b** rotational compliant displacements ϕ

$$M_2 = 0 \tag{8.21}$$

From the theory of elasticity, we can formulate the translational displacements of point A and point B, which are δ_1 and δ_2, respectively:

$$\delta_1 = \frac{\ell_1^3}{3E_1 I_1} F_1 + \frac{\ell_1^2}{3E_1 I_1} M_1 \tag{8.22}$$

$$\delta_2 = \frac{\ell_2^3}{3E_2 I_2} F_2 + \frac{\ell_2^2}{3E_2 I_2} M_2 \tag{8.23}$$

Also, from the theory of elasticity we can formulate the rotational displacement of point A and point B, which are ϕ_1 and ϕ_2, respectively:

$$\phi_1 = \frac{\ell_1^2}{3E_1 I_1} F_1 + \frac{\ell_1}{E_1 I_1} M_1 \tag{8.24}$$

$$\phi_2 = \frac{\ell_2^2}{2E_2 I_2} F_2 + \frac{\ell_2}{E_2 I_2} M_2 \tag{8.25}$$

We can formulate ϕ_1 and ϕ_2 in terms of δ_1 and δ_2 for the sake of finding the Jacobian of elastic deformations J_ℓ

$$\phi_1 = \frac{3}{2\ell_1}\delta_1 + \frac{3E_2I_2\ell_1\ell_2}{2E_1I_1\left(2\ell_2^3 + 3\ell_2^2\right)}\delta_2 \tag{8.26}$$

$$\phi_2 = \frac{3\left(\ell_2^2 + 2\ell_2\right)}{2\ell_2^3 + 3\ell_2^2}\delta_2 \tag{8.27}$$

The linear and angular movements of point B can be expressed by the vector f_T

$$f_T = \begin{Bmatrix} x_B \\ y_B \\ \theta_T \end{Bmatrix} \tag{8.28}$$

where

$$\theta_T = \theta_1 + \phi_1 + \theta_2 + \phi_2 \tag{8.29}$$

To find the compliance matrix of the links, we shall consider all the elastic deformations experienced by the two links:

$$C_\ell = J_\ell K_\ell^{-1} J_\ell^T \tag{8.30}$$

$$K_\ell = \text{diag}(k_1, k_2) \tag{8.31}$$

$$x_\ell = \begin{bmatrix} \delta_1 \\ \delta_2 \end{bmatrix} \tag{8.32}$$

$$J_\ell = \frac{\partial f_T}{\partial x_\ell} \tag{8.33}$$

$$J_\ell = \begin{bmatrix} J_{\ell 11} & J_{\ell 12} \\ J_{\ell 21} & J_{\ell 22} \\ J_{\ell 31} & J_{\ell 32} \end{bmatrix} \tag{8.34}$$

$$J_{\ell 11} = -\sin(\theta_1) - \frac{3\ell_2\sin(\theta_{\text{aux}})}{2\ell_1} - \frac{3\delta_2\cos(\theta_{\text{aux}})}{2\ell_1} \tag{8.35}$$

$$J_{\ell 12} = -\frac{3E_2I_2\ell_1\sin(\theta_{\text{aux}})}{4\ell_2E_1I_1} - \sin(\theta_{\text{aux}}) - \frac{3\delta_2E_2I_2\ell_1\cos(\theta_{\text{aux}})}{4\ell_2^2E_1I_1} \tag{8.36}$$

$$J_{\ell 21} = \cos(\theta_1) + \frac{3\ell_2\cos(\theta_{\text{aux}})}{2\ell_1} - \frac{3\delta_2\sin(\theta_{\text{aux}})}{2\ell_1} \tag{8.37}$$

$$J_{\ell 22} = -\frac{3E_2I_2\ell_1\cos(\theta_{\text{aux}})}{4\ell_2E_1I_1} + \cos(\theta_{\text{aux}}) - \frac{3\delta_2E_2I_2\ell_1\sin(\theta_{\text{aux}})}{4\ell_2^2E_1I_1} \tag{8.38}$$

$$J_{\ell 31} = \frac{3}{2\ell_1} \tag{8.39}$$

$$J_{\ell 32} = -\frac{3E_2 I_2 \ell_1}{4\ell_2^2 E_1 I_1} + \frac{3}{2\ell_2} \tag{8.40}$$

$$\theta_{\text{aux}} = \theta_1 + \frac{3\delta_1}{2\ell_1} + \frac{3\delta_2 E_2 I_2 \ell_1}{4\ell_2^2 E_1 I_1} + \theta_2 \tag{8.41}$$

The strain energy of this system is described by Komatsu (1990) as follows:

$$U = \frac{k_1 \delta_1^2 + k_{12}\delta_1\delta_2 + k_2\delta_2^2}{2} \tag{8.42}$$

while the strain energy of the beam is described by Beer et al. (2009):

$$U = \frac{1}{2E_1 I_1}\int_0^{\ell_1} (F_1 x_1 + M_1)^2 \mathrm{d}x_1 + \frac{1}{2E_2 I_2}\int_0^{\ell_2} (F_2 x_2 + M_2)^2 \mathrm{d}x_2. \tag{8.43}$$

By solving Eqs. (8.42) and (8.43), we can find the stiffness components

$$k_1 = \frac{3E_1 I_1}{\ell_1^3} \tag{8.44}$$

$$k_{12} = 0 \tag{8.45}$$

$$k_2 = \frac{9\ell_1\ell_2 E_2^2 I_2^2 (4\ell_1 - 4\ell_1\cos\theta_2 + \ell_2) + 12E_1 I_1 E_2 I_2 \ell_2^3}{4E_1 I_1 \ell_2^6} \tag{8.46}$$

By substitution of all of the components in (8.30), we get C_ℓ
The compliance matrix due to compliant joints is calculated as follows:

$$C_{jt} = J_{jt} K_{jt}^{-1} J_{jt}^T, \tag{8.47}$$

where $K_{jt} = \text{diag}\left(k_{jt1}, k_{jt2}\right)$,

$$J_{jt} = \frac{\partial f_T}{\partial q_i}; \quad q = \begin{Bmatrix} \theta_1 \\ \theta_2 \end{Bmatrix}. \tag{8.48}$$

$$J_{\text{art}} = \begin{bmatrix} -\ell_1\sin\theta_1 - \delta_1\cos\theta_1 - \ell_2\sin\theta_{\text{aux}} - \delta_2\cos\theta_{\text{aux}} & -\ell_2\sin\theta_{\text{aux}} - \delta_2\cos\theta_{\text{aux}} \\ \ell_1\cos\theta_1 - \delta_1\sin\theta_1 + \ell_2\cos\theta_{\text{aux}} - \delta_2\sin\theta_{\text{aux}} & \ell_2\cos\theta_{\text{aux}} - \delta_2\sin\theta_{\text{aux}} \end{bmatrix} \tag{8.49}$$

We can find the stiffness coefficients k_{jt1} and k_{jt2} from experimental tests or directly from the manufacturer's catalogue. Finally, by summation of the two compliance matrices we find the total compliance of the system and global stiffness matrix

$$C_T = C_\ell + C_{jt}$$
$$K = C_T^{-1} \tag{8.50}$$

8.5 Matrix Structural Analysis

In this section, we will learn how to calculate the stiffness matrix using matrix structural analysis (MSA), which is also called the displacement method or direct stiffness method (DSM) (Júnior et al. 2016). MSA is the inverse of the finite element method (FEM), and it breaks up a complex system into a discrete set of simple components. The stiffness of these simple elements can be easily determined. Manipulators are composed of joints and links, the stiffness of each of them can be calculated individually, and then the stiffness of the whole manipulator can be determined by superposition of all of the stiffness matrices of joints and links. When an external wrench is applied to a manipulator, each link and joint examine rotational and translational compliants. The stiffness matrix is greatly affected by the boundary conditions and the configuration of the mechanical system. By using a set of rules, we can fit all the matrices (building blocks derived from the theory of elasticity) together to find the matrix which represents the stiffness of the whole robot. Then by using Eq. (8.2), we can find the corresponding compliance due to external loading.

8.5.1 *Stiffness of Link and Joint*

Stiffness of the joint can be represented by the following equation:

$$K_{\text{joint}} = \begin{bmatrix} k_c & -k_c \\ -k_c & k_c \end{bmatrix} \tag{8.51}$$

where $k_c = \text{diag}\left(k_{tx}\ k_{ty}\ k_{tz}\ k_{rx}\ k_{ry}\ k_{rz}\right)$, $\left(k_{tx}\ k_{ty}\ k_{tz}\right)^T$ are the translational stiffnesses in the x-, y-, z-axes while $\left(k_{rx}\ k_{ry}\ k_{rz}\right)^T$ are the rotational stiffnesses in the three axes. Stiffness components are usually determined from the manufacturer's catalogue or by experimental tests. Considering Figs. 8.3 and 8.4, where a link with uniform cross-section is assumed, the stiffness of the link can be found by

$$K_\ell = \begin{bmatrix} k_{\ell j} & -k_{\ell j} \\ -k_{\ell j} & k_{\ell j} \end{bmatrix} \tag{8.52}$$

where k_ℓ is defined by the 6×6 matrix shown in Eq. (8.3) and j is the index of the link in the manipulator. Before the superposition of the whole matrix, each joint stiffness matrix and link stiffness matrix should be represented to the inertial frame using the transformation matrices

$$K_{\text{joint}}^{e} = R_i^o K_{\text{joint}} R_i^{0T} \tag{8.53}$$

$$K_{\ell}^{e} = R_i^o K_{\ell} R_i^{0T} \tag{8.54}$$

Therefore, we can rewrite Eqs. (8.53) and (8.54) as

$$K_{\text{joint}}^{e} = \begin{bmatrix} R_i^0 & 0 & 0 & 0 \\ 0 & R_i^0 & 0 & 0 \\ 0 & 0 & R_i^0 & 0 \\ 0 & 0 & 0 & R_i^0 \end{bmatrix} \begin{bmatrix} k_c & -k_c \\ k_c & -k_c \end{bmatrix} \begin{bmatrix} R_i^{0T} & 0 & 0 & 0 \\ 0 & R_i^{0T} & 0 & 0 \\ 0 & 0 & R_i^{0T} & 0 \\ 0 & 0 & 0 & R_i^{0T} \end{bmatrix} \tag{8.55}$$

$$K_{\ell}^{e} = \begin{bmatrix} R_i^0 & 0 & 0 & 0 \\ 0 & R_i^0 & 0 & 0 \\ 0 & 0 & R_i^0 & 0 \\ 0 & 0 & 0 & R_i^0 \end{bmatrix} \begin{bmatrix} k_{\ell} & -k_{\ell} \\ k_{\ell} & -k_{\ell} \end{bmatrix} \begin{bmatrix} R_i^{0T} & 0 & 0 & 0 \\ 0 & R_i^{0T} & 0 & 0 \\ 0 & 0 & R_i^{0T} & 0 \\ 0 & 0 & 0 & R_i^{0T} \end{bmatrix} \tag{8.56}$$

After all matrices are represented to the base frame, MSA can be applied to find the stiffness of the structure of the manipulator. The methodology described for MSA assumes that the whole structure is free. In other words, the global stiffness matrix for the system will be singular and cannot be inverted. To solve this problem, boundary conditions must be applied to this matrix to convert it to an invertible version to find the displacements:

$$\Delta x = K^{-1} F \tag{8.57}$$

8.5.2 *Methodology*

Assume we have to find the stiffness matrix of the structure shown in Fig. 8.7a, which consists of two beams connected together utilizing spherical joints. Each physical connection and point of force effect should be represented as a node. Figure 8.7b shows the possible node on the structure, where node 2 is the point of the effect of force F_y, which splits the horizontal beam into two segments. Now, the model is composed of segment 1–2, segment 2–3, joint 3–4, and segment 4–5. Note that the boundary conditions for this system are that node 1 and node 5 are fixed with zero translational and rotational displacements. It is worth mentioning that element 3–4, which represents the spherical joint, is only a representative scheme where nodes 2 and 3 are considered to form the joint. Recall Fig. 8.4: Each node has 6DOF $\left(\delta_{xi}\ \delta_{yi}\ \delta_{zi}\ \phi_{xi}\ \phi_{yi}\ \phi_{zi}\right)^T$, and each element (joint and link) has 12DOF. Each

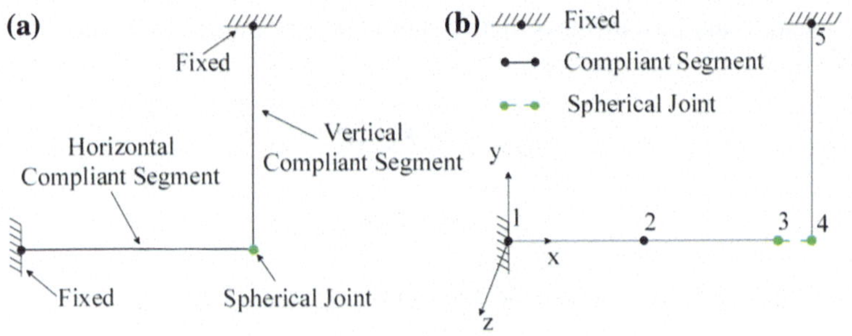

Fig. 8.7 Structure of two beams connected by spherical joint

element stiffness matrix, Eqs. (8.51) and (8.52), should be represented to the inertial frame using transformation matrices as in Eqs. (8.55) and (8.56). The similarity transformation in Eqs. (8.55) and (8.56) is the reason that the global stiffness matrix obtained by MSA is configuration dependent. Table 8.1 shows the degree of freedom of each of the five nodes. Each DOF should have a different identification number, and it is preferable that they be in an ascendant manner. A connectivity matrix is formed in which each row represents the degrees of freedom of each element. Thus, we have four rows, each row with twelve indices (12DOF represent 6DOF for node *i*). The connectivity matrix shows that we have 30DOF for the proposed structure. Consequently, the global stiffness matrix of the whole structure should have 30×30 dimensions. The connectivity matrix provides us with information about how the DOFs relate to each other.

$$\text{connec} = \begin{bmatrix} 1 & 2 & 3 & 4 & 5 & 6 & 7 & 8 & 9 & 10 & 11 & 12 \\ 7 & 8 & 9 & 10 & 11 & 12 & 13 & 14 & 15 & 16 & 17 & 18 \\ 13 & 14 & 15 & 16 & 17 & 18 & 19 & 20 & 21 & 22 & 23 & 24 \\ 19 & 20 & 21 & 22 & 23 & 24 & 25 & 26 & 27 & 28 & 29 & 30 \end{bmatrix} \begin{matrix} \text{segment } 1-2 \\ \text{segment } 2-3 \\ \text{joint } 3-4 \\ \text{segment } 4-5 \end{matrix} \tag{8.58}$$

As a start, we can define the global matrix with 30×30 zero values as shown in Fig. 8.8.

Table 8.1 Degrees of freedom of each node in the structure

Compliant displacement	Node $i = 1$	Node $i = 2$	Node $i = 3$	Node $i = 4$	Node $i = 5$
δ_{xi}	1	7	13	19	25
δ_{yi}	2	8	14	20	26
δ_{zi}	3	9	15	21	27
ϕ_{xi}	4	10	16	22	28
ϕ_{yi}	5	11	17	23	29
ϕ_{zi}	6	12	18	24	30

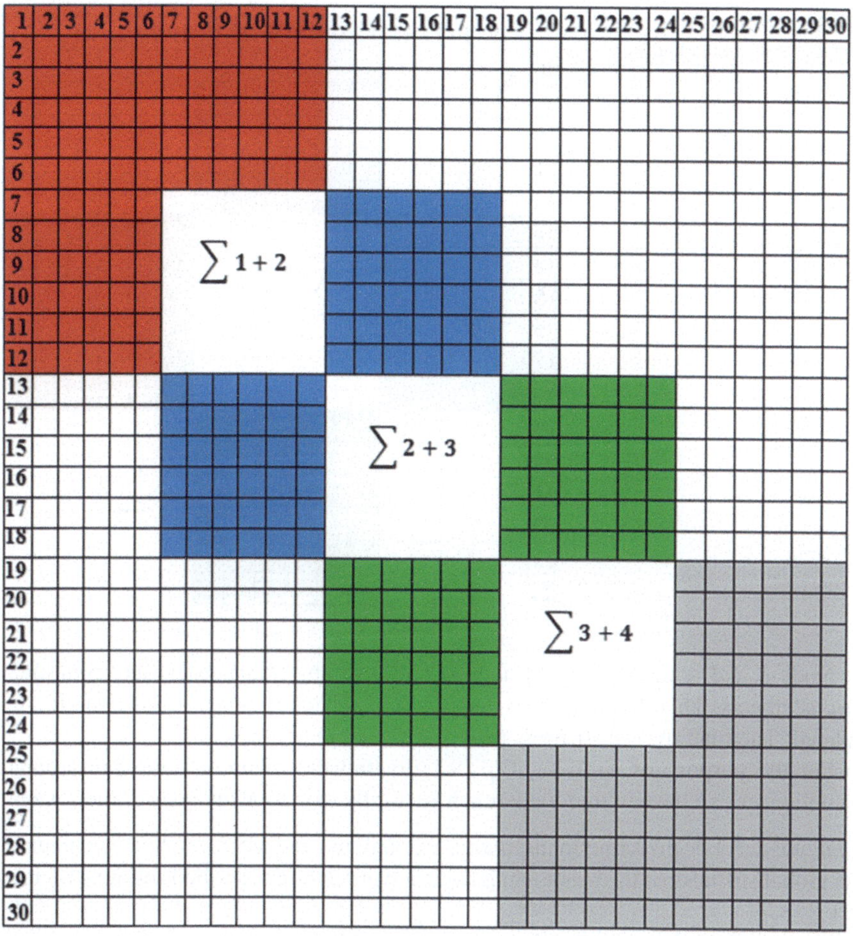

Fig. 8.8 Assembling the structure stiffness matrix

For segment 1–2, the DOFs are identified by numbers 1–12. From the index (1,1) to (12,12) in the global matrix, elements should be replaced by the stiffness matrix shown in Eq. (8.56). Segment 2–3 is defined by DOFs 7–18 in the connectivity matrix, and from the index (7,7) to (18,18) we have replaced elements by the beam matrix (8.56). In the same way, indices from (13,13) to (24,24) should be replaced by the stiffness of the joint (3–4) defined by DOFs 13–24, which is shown in Eq. (8.55). The stiffness of the final segment 4–5, which is charaterized by DOFs 19–30 in the connectivity matrix should be put instead of elements (19,19) to (30,30) in the global matrix. The overlapping regions should be calculated as the summation of overlapped components. Once again, the resulting global matrix is singular, and boundary conditions should be applied.

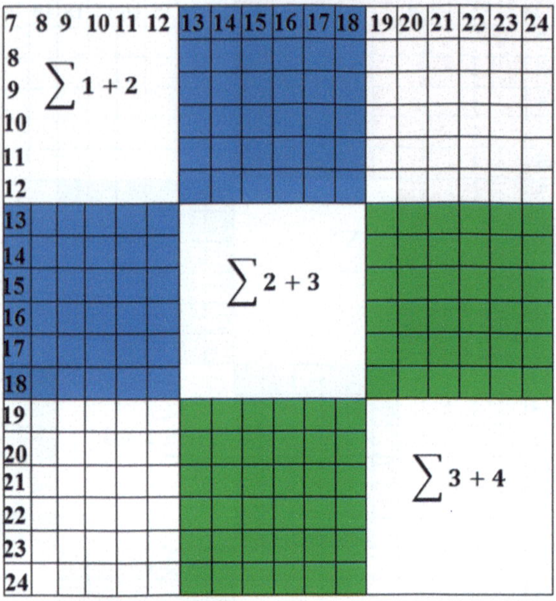

Fig. 8.9 Final overall stiffness matrix obtained by MSA

Node 1 and node 5 are fixed and do not have any translational and rotational displacements. Therefore, all rows and columns related to nodes 1 and 5 should be deleted. The final overall stiffness matrix is shown in Fig. 8.9.

For the purpose of applying Eq. (8.2) to find the rotational and translational compliances, we have to find the wrench vector by structural analysis of the system.

Example 8.2 For the same manipulator, we may use the MSA method to calculate the global stiffness of the robot using the same procedure described for the structure in Fig. 8.7.

First of all, we have to identify each element in this mechanical system, as shown in Fig. 8.10. For this example, the robot consists of 5 segments and 6 nodes starting from the ground where node 1 is attached. Node 1 is fixed to the ground; they have zero translational and rotational displacements, and also the inertial frame is attached to this point. Segment 1–2 is considered rigid, and it is assumed that it has a rigidity that is ten times greater than other segments. We have to identify the numbering DOFs of each node, as described in Table 8.2.

The components of the joint stiffness matrix in Eq. (8.51) can be obtained from experimental tests or the manufacturer's catalogue. The link is considered with a circular cross section, and by neglecting shear forces, the link stiffness matrix is expressed in Eq. (8.52). The stiffness of segments 1–2, 3–4, and 5–6 is represented by (8.3), but we should note that the G value of segment 1–2 is equal to ten times that of the other segments because we have assumed it to be a rigid beam. Segments 2–3 and 4–5 are represented by Eq. (8.51), where they are joint elements. Before building the global matrix, all five matrices should be transformed into the inertial frame at node 1 using Eqs. (8.55) and (8.56).

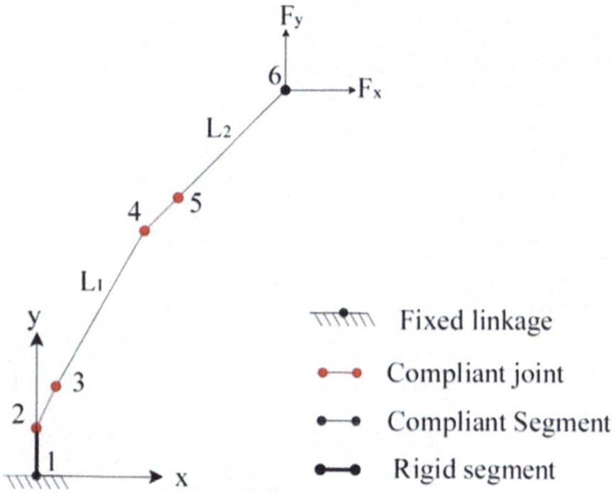

Fig. 8.10 Schematic diagram of all elements of the 2DOF manipulator

Table 8.2 Degrees of freedom of each node in the 2DOF manipulator

Compliant displacement	Node i = 1	Node i = 2	Node i = 3	Node i = 4	Node i = 5	Node i = 6
δ_{xi}	1	7	13	19	25	31
δ_{yi}	2	8	14	20	26	32
δ_{zi}	3	9	15	21	27	33
ϕ_{xi}	4	10	16	22	28	34
ϕ_{yi}	5	11	17	23	29	35
ϕ_{zi}	6	12	18	24	30	36

The following is the connectivity matrix, where five rows are representing the five segments of the model.

$$\text{connec} = \begin{bmatrix} 1 & 2 & 3 & 4 & 5 & 6 & 7 & 8 & 9 & 10 & 11 & 12 \\ 7 & 8 & 9 & 10 & 11 & 12 & 13 & 14 & 15 & 16 & 17 & 18 \\ 13 & 14 & 15 & 16 & 17 & 18 & 19 & 20 & 21 & 22 & 23 & 24 \\ 19 & 20 & 21 & 22 & 23 & 24 & 25 & 26 & 27 & 28 & 29 & 30 \\ 25 & 26 & 27 & 28 & 29 & 30 & 31 & 32 & 33 & 34 & 35 & 36 \end{bmatrix} \begin{matrix} \text{seg. } 1-2 \\ \text{seg. } 2-3 \\ \text{seg. } 3-4 \\ \text{seg } 4-5 \\ \text{seg } 5-6 \end{matrix} \tag{8.59}$$

The dimensions of the global stiffness matrix are $6n = 36 \times 36$ where n is the number of nodes, as shown in Fig. 8.11. By using the information of the connectivity matrix in Eq. (8.57), we insert the building blocks of the 6×6 matrices of joints and links into the global 36×36. Matrices of segments 1–2, 2–3, 3–4, 4–5, and 5–6 are inserted to replace elements of the global matrix starting from (1,1), (7,7), (13,13), (19,19), and (25,25), respectively. By applying the boundary condition, which is node 1 if fixed, we delete all rows and columns related to the degree of freedom of node 1. Figure 8.12 reveals the final global stiffness matrix of the robot which is

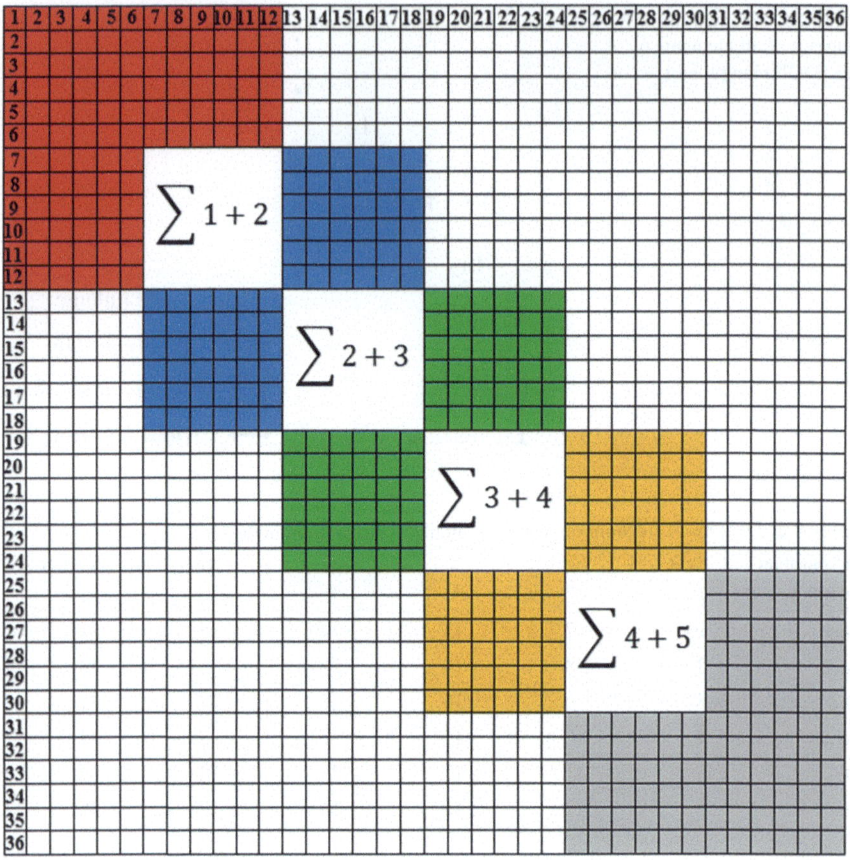

Fig. 8.11 Assembling the overall stiffness matrix of the 2DOF robot

invertible 30×30 matrix.

8.6 Summary

In this chapter, the elastic deformation of robot manipulator components is conducted using different methods. Calculation of the compliant displacements is necessary during run-time because it provides us with proper predictions about the error in positioning the end-effector of the robot. These elastic deformations can be provided to the control system to correct the position of the end-effector by adjusting the set of joint angles. On the other hand, we have to keep our robot works in the elastic zone by choosing appropriate loading conditions.

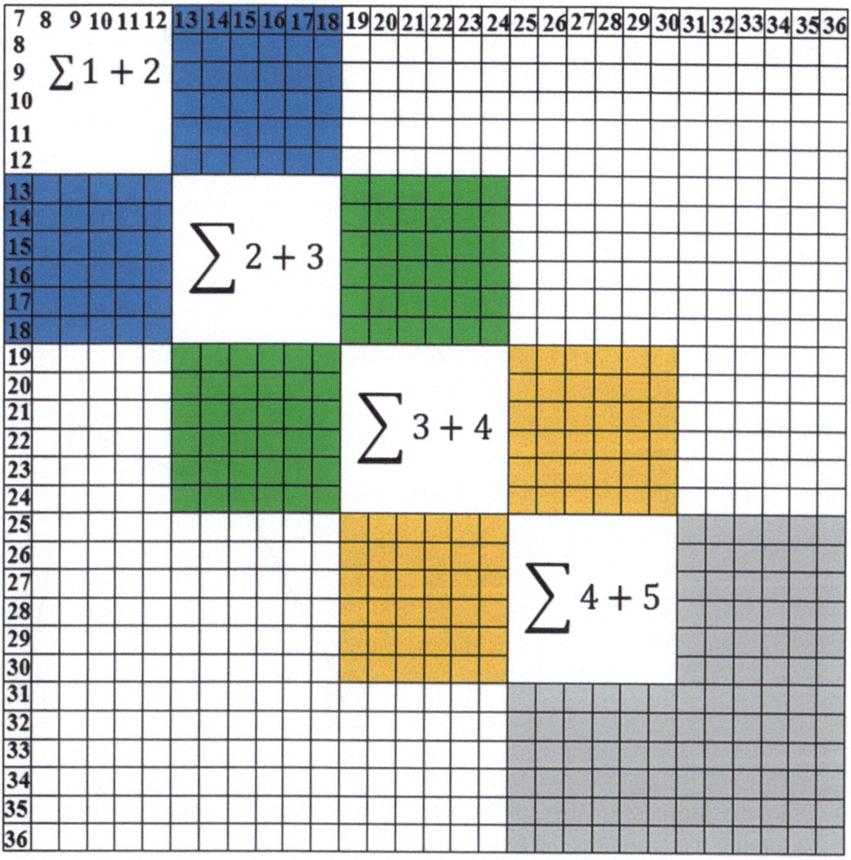

Fig. 8.12 Final global stiffness matrix for 2DOF manipulator

Exercises

8.1 Assume that links of the manipulator shown in Fig. 8.13 have the same stiffness matrix which is represented in Eq. (8.3) and determine the whole stiffness matrix of the manipulator. Assume any needed data.

8.2 Solve Example 8.2 with numerical data of your choice.

8.3 Assume that the planar robot in Example 8.2 carries 100 kg, what is the best choice.

a. Choose a circular cross-sectional area beam of 300-mm outer diameter and 10-mm thickness.

b. Choose a rectangular cross-sectional area with a 300-mm inner diameter and 10-mm thickness.

Assume all the required data.

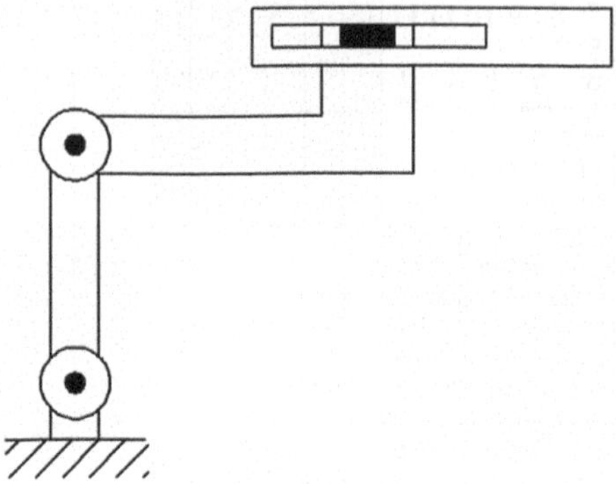

Fig. 8.13 Three-link manipulator

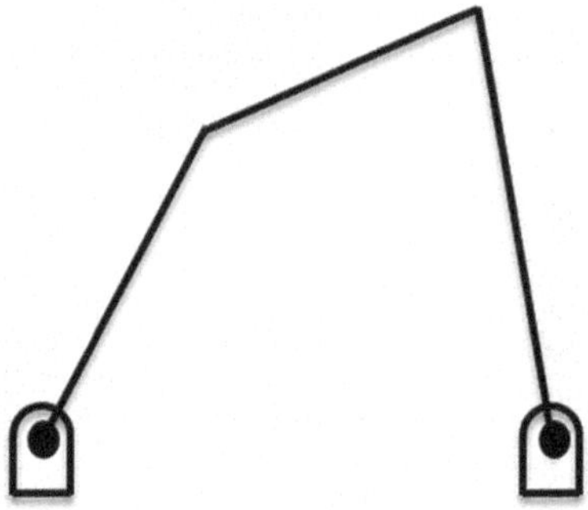

Fig. 8.14 Four-link mechanism

8.4 Find the stiffness matrix of the four-link mechanism shown in Fig. 8.14 assuming the ground is the fourth link.

References

Beer F, Johnston E Jr, DeWolf J, Mazurek D (2009) Mechanics of materials, 5th edn. McGraw-Hill, New York, NY

Ghafil HN, Jármai K (2019) Kinematic-based structural optimization of robots. Pollack Periodica 14(3):213–222

Gonçalves RS, Carbone G, Carvalho JCM, Ceccarelli M (2016) A comparison of stiffness analysis methods for robotic systems. Int J Mech Control 17(2):35–58

Júnior GS, Carvalho JCM, Gonçalves RS (2016) Stiffness analysis of multibody systems using matrix structural analysis—MSA. Robotica 34(10):2368–2385

Komatsu T (1990) Compliance control for a two-link flexible manipulator. Trans JSME (C) 56(530):2642–2648

Tsai L-W (1999) Robot analysis: the mechanics of serial and parallel manipulators. Wiley, Hoboken

Yoon W-K, Suehiro T, Tsumaki Y, Uchiyama M (2004) Stiffness analysis and design of a compact modified delta parallel mechanism. Robotica 22(4):463–475

Chapter 9
Kinematic Synthesis

9.1 Introduction

In this chapter, we will study the concept of the kinematic synthesis of robot manipulators which is an optimization problem searches for the best topology and dimensionality of a mechanism. Designers have been achieving this operation by hand, and the results depend on their skills and ingenuity. However, since the middle of the last century, many researchers have been trying to do that in a systematic and automated manner. Kinematic synthesis is time-consuming, and there is no robust approach to find kinematic synthesis to meet complex tasks that can be achieved by a robot. Two strategies will be explained during this chapter which are genetic strategy and planning strategy. Kinematic synthesis consists of two characteristics, type synthesis and dimensional synthesis.

9.2 Type Synthesis

Also called topology or topological synthesis, it is a dimensionless problem that searches for the best robot type which can be done the work. In other words, how many links in the manipulator, how many joints in the manipulator, how these links and joints connect with each other. Thus, type synthesis describes the spatial characteristics of the robot, regardless of the dimensions of the components.

9.3 Dimensional Synthesis

Dimensionality or dimensional synthesis is to find the best dimensions of the robot components to perform the desired trajectory or motion task.

H. N. Ghafil and K. Jármai (ed.), *Optimization for Robot Modelling with MATLAB*,
https://doi.org/10.1007/978-3-030-40410-9_9

9.4 Genetic Algorithms

In 1975, Holland (Mühlenbein 1997) has introduced the concept of a genetic algorithm (GA) as a solver which can be used to solve optimization problems. Although numbers can be represented as strings of real or binary numbers, we will employ the binary representation, as shown in Fig. 9.1.

Usually, the initial population of the GA is taken randomly and then improved during a specific time or number of iterations. Three main strategies can be distinguished; the first one is the selection which is the operation of selecting parents among the population, the selection in probabilistic operation which gives a great chance to the best elements in the population to be parents. Consider Fig. 9.1, we can select, for instance, 41 and 78 to be parents. After transforming these decimal numbers to the binary form, we apply the second strategy, which is crossover. In crossover a block of genes (0,1) from one parent can be replaced with another block of genes from another parent; in our example, we could replace block *b* by *d* and vice versa. The result is two different binary strings that can be exposed to the mutation operation which is replacing a specific gene with different value, for instance, the gene highlighted in grey is replaced by 0 instead of 1. Crossover and mutation are random operations, and there are a lot of methods on how to do that.

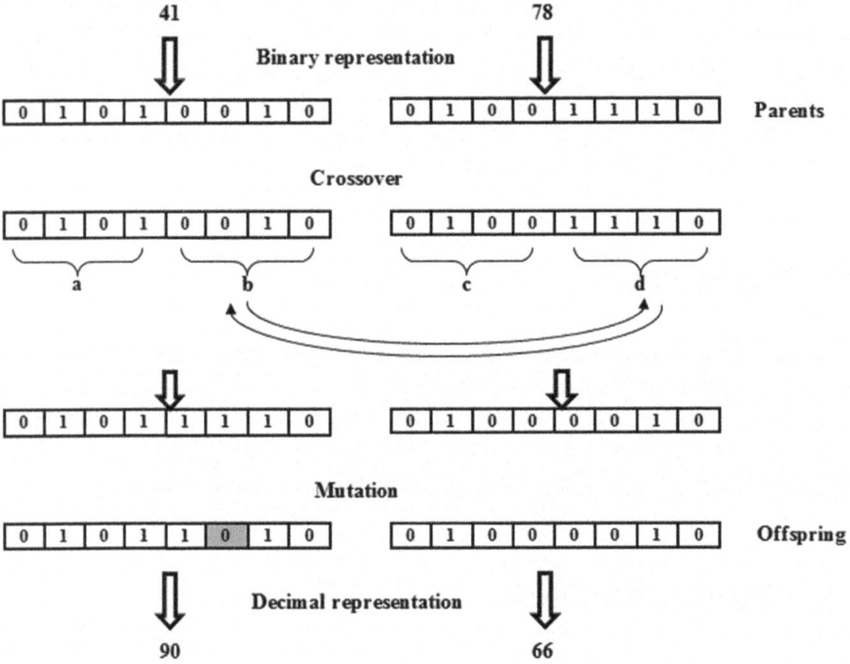

Fig. 9.1 Genetic algorithm

9.5 Genetic Representation of a Mechanism

In this section, we will explain how to represent a mechanism by genetic operators with an example given by (Oliva and Goodman 2010) and shown in Fig. 9.2. Genetically, this mechanism may be represented by a chromosome of seven genes each one corresponds to a physical component. Two genes for x and y coordinates of the pivot point a, Two genes for x and y coordinates of the pivot point d, gene for link length ab, a gene for link length bc, and a gene for link length cd that are shown in Table 9.1. In the beginning, these seven genes initiated randomly with random numbers, by assuming that the objective is to meet a given motion, GA keeps iterated until it finds the proper values of the genes that enable the 4-bar mechanism to perform the given motion.

This genetic representation of the 4-bar linkage is an example of how we can represent a mechanism as a genetic parameter that can be optimized during the run-time of the genetic algorithm. Thus, in this example, we search for optimum dimensionality for the mechanism. In other words, we have a given topology or type synthesis, and the optimization occurs to find the dimensional synthesis. In the following sections, we will present how to find the topology using two strategies; the first strategy depends on graph theory formulation, and it is an evolutionary strategy. The second one is the planning strategy which depends on path planning

Fig. 9.2 Four-bar mechanism

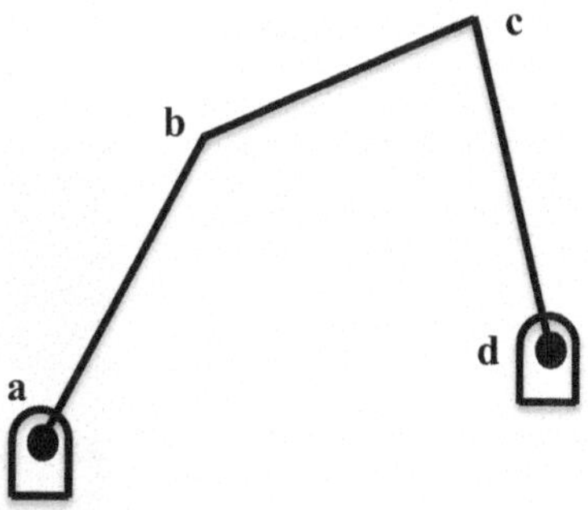

Table 9.1 Chromosome of the 4-bar mechanism

Gene No.	Physical description
1	Pivot a, x-coordinate
2	Pivot a, y-coordinate
3	Pivot d, x-coordinate
4	Pivot d, y-coordinate
5	Link length ab
6	Link length bc
7	Link length cd

discipline. We have dealt with path planning in Chap. 6, so it is good to start with some terminologies about graph theory which is the backbone of evolutionary methods.

9.6 Graph Theory

The first form of this theory was established by Leonhard Euler (Thiele 2005) while trying to solve the seven bridges of Königsberg problem, see Fig. 9.3, this problem had asked for a path that can cross over all the seven bridges such that each one should be crossed once.

Graph theory (West 2001; Bollobás 2013) has some terminologies that can be described briefly in this text to give background for subsequent sections.

9.6.1 Graph

A collection of sets of vertices V and edges E is called graph and can be described by the triple (V, E, g). V is always nonempty set while E could be an empty set. g is mapping called incidence mapping, and for each edge, $e \in E$ there is subset $g(e) = \{u, v\}$ where u and v are vertices. Figure 9.4 explains an example of a graph with four vertices v_1, v_2, v_3, and v_4 as well as four edges e_1, e_2, e_3, and e_4.

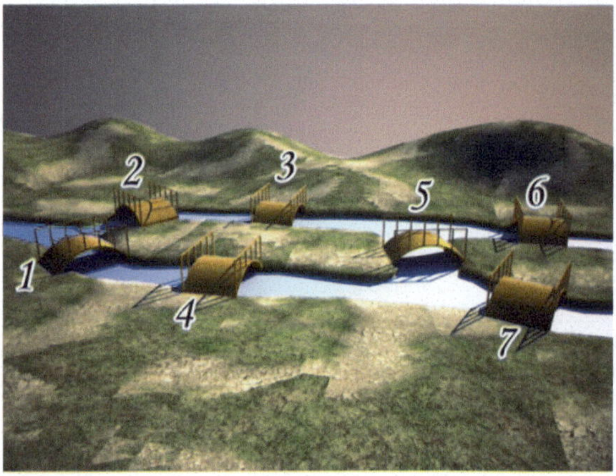

Fig. 9.3 Description of the seven bridges of Königsberg

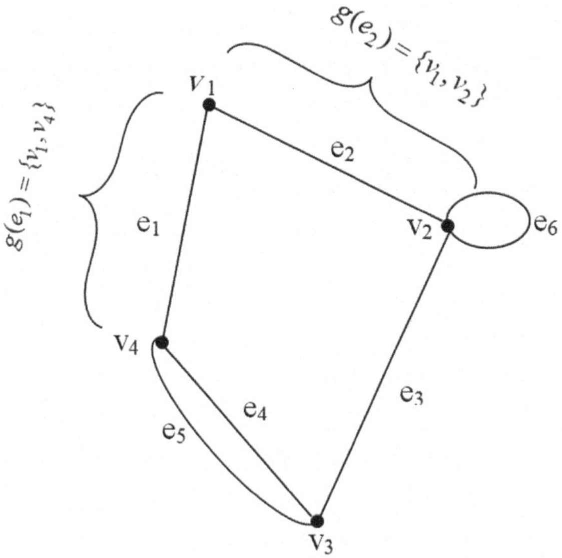

Fig. 9.4 Graph representation

9.6.2 Incidence

Consider Fig. 9.2, every edge is said to be incident with the vertices that it connects, so edge e_2 is incident with vertices v_1 and v_2. Also, the vertex is said to be incident vertex if it belongs to one or more edges, so v_1 is incident with edges e_1 and e_2.

9.6.3 Adjacent Vertices

Vertices are said to be adjacent if the same edge connects them, for example, v_1 is adjacent to v_2 and v_4.

9.6.4 Adjacent Edge

The edge is said to be adjacent to another if they have a common vertex, for example, e_3 and e_4 are adjacent edges.

9.6.5 Self-loop

An edge has the same end vertices, for example, e_6 in Fig. 9.4.

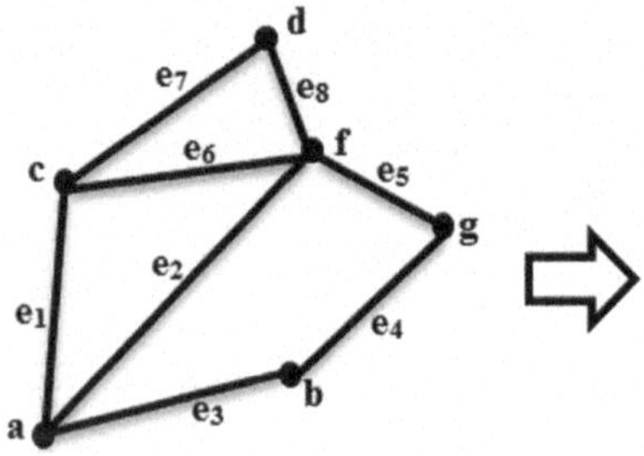

IM=

	e_1	e_2	e_3	e_4	e_5	e_6	e_7	e_8
a	1	1	1	0	0	0	0	0
b	0	0	1	1	0	0	0	0
c	1	0	0	0	0	1	1	0
d	0	0	0	0	0	0	1	1
f	0	1	0	0	1	1	0	1
g	0	0	0	1	1	0	0	0

Fig. 9.5 Graph and its incidence matrix

9.6.6 *Parallel Edge*

An edge shares its end vertices with another edge, e_5 in Fig. 9.4.

9.6.7 *Incidence Matrix*

Incidence matrix MI of a graph is a matrix whose elements (v, e) are 0 or 1, each vertex has a row, and each edge has a column. The element $(v, e) = 1$ if and only if the vertex is incident with an edge, and it is zero if not, for instance, Fig. 9.5 shows a graph with its IM matrix.

IM gives a unique binary representation of a specific graph regardless of the dimensions of the edges consisting the graph. The inverse problem is correct; we can reach a unique graph in Fig. 9.5 if we find optimum IM starting from a randomly generated matrix. This is what Liu approach employs to find optimum type synthesis of a mechanism.

9.7 Liu Approach

Liu and McPhee (2007) have supposed that every linkage can be represented by an incidence matrix. Every single link can be a binary link, ternary link, or quaternary link, and in this book, we will not deal with quaternary link. The binary link consists of single-segment connects two nodes while the ternary link consists of three segments connects three nodes, as shown in Fig. 9.6.

The incidence matrix IM representing a mechanism taking a binary form, as shown in Fig. 9.7, this form is convenient for genetic algorithm which represents numbers as a string of binary numbers. Consequently, Liu approach uses genetic algorithm to optimize randomly generated IM to find feasible topology solution.

The candidate solution in Liu approach is an extendable dimension IM which can be crossed over and mutated with others, as shown in Figs. 9.8 and 9.9.

Link type	Diagram	Parameters	Parameter number
Binary link	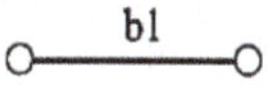	b1	1
Ternary link	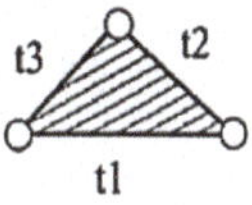	t1, t2, t3	3

Fig. 9.6 Geometric parameters of standard links

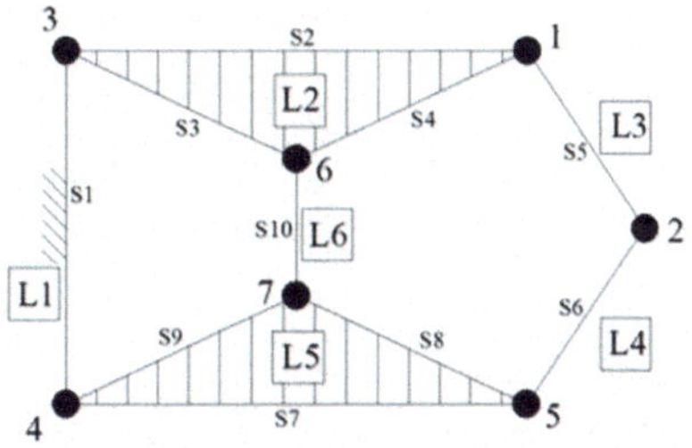

	L1	L2	L3	L4	L5	L6
1	0	1	1	0	0	0
2	0	0	1	1	0	0
3	1	1	0	0	0	0
4	1	0	0	0	1	0
5	0	0	0	1	1	0
6	0	1	0	0	0	1
7	0	0	0	0	1	1

Fig. 9.7 Representation of the 6-bar linkage by a unique IM

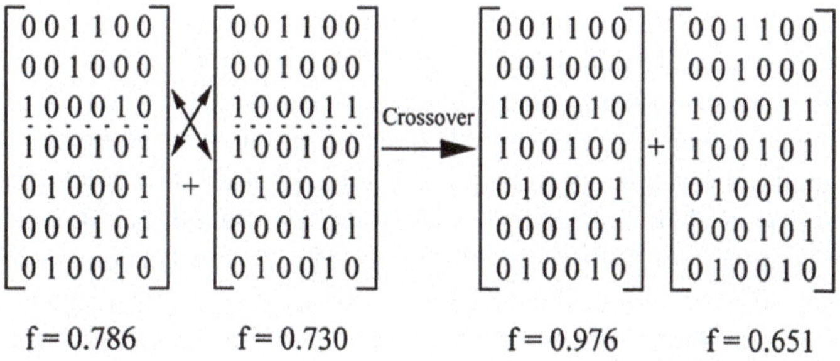

Fig. 9.8 A crossover between two IMs

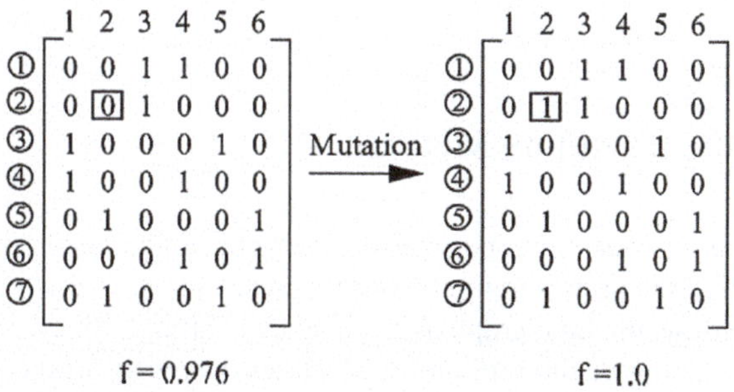

Fig. 9.9 Mutation on single IM

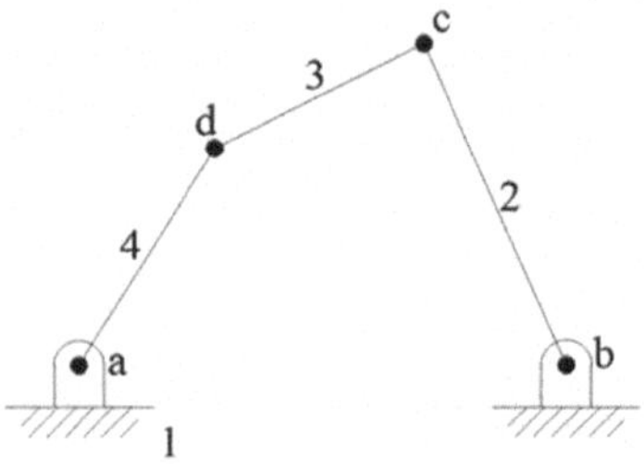

	1	*2*	*3*	*4*
a	1	0	0	1
b	1	1	0	0
c	0	1	1	0
d	0	0	1	1

Fig. 9.10 4-bar planar linkage

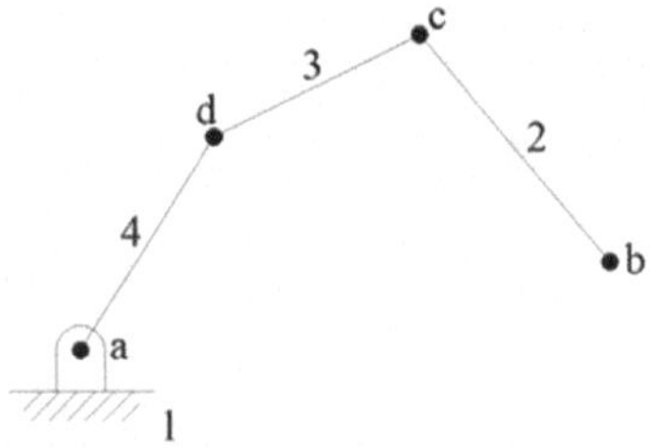

	1	*2*	*3*	*4*
a	1	0	0	1
b	0	1	0	0
c	0	1	1	0
d	0	0	1	1

Fig. 9.11 3-bar open-chain planar manipulator

The objective function could finite elements model which can be solved for strength, and this will be the fitness of the solution. There are optimization softwares for a mechanical design where the user formulates the objective, and the software takes care of the optimization process. For example, the genetic solutions in Liu approach are evaluated by KIDDS package while (Oliva and Goodman 2010) have used HEEDS Optimization Software. This approach was intended for a closed-chain planar mechanism, but it can be modified for open-chain mechanisms. Consider Fig. 9.10 which is 4-bar linkage and represented by 4×4 IM; this mechanism can be modified to be open-chain linkage by freeing node (b) from link 1 as shown Fig. 9.11.

This can be done by setting a condition that IM (2,1) always equal to zero. This approach and all other graph theory-based methods take a given trajectory as an input and return the proposed mechanism as an output. This is an essential difference between these methods and planning method which generates its proper trajectory according to the given problem. The input for planning method is base point and task point of the manipulator as well as the obstacles in the workspace.

9.8 Planning Method

The A* algorithm is suggested to be the path planner that generates the initial graph in a free or constrained working area. Choosing the A* algorithm is not arbitrary; it returns the final graph as a sequence of straight lines, and this is the basic idea of the planning method. Figure 9.12 shows the results of the path planning process, where the dark circles represent obstacles within the working area. Figure 9.12 clearly shows that the path of A* algorithm is a combination of straight lines connecting

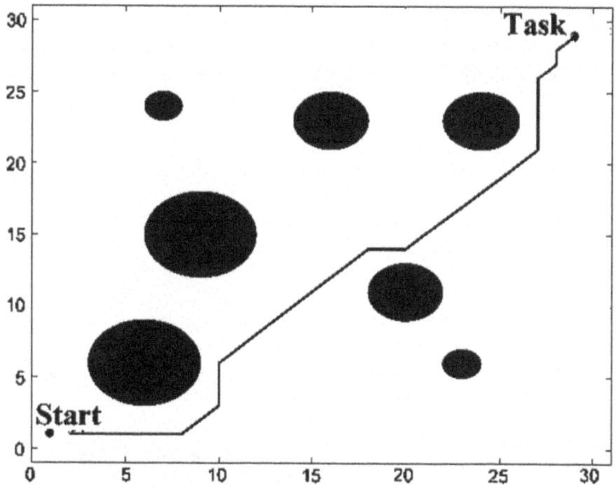

Fig. 9.12 Planned path between two given points

the start and task points. Other planners that depend on cubic spline curves are not appropriate for the planning method because these planners usually return a graph in the form of a continuous curve with high order. Planning method needs to discontinue the relationship between base and task points, and this linear discontinues formulation can be provided by A* algorithm.

The points that make up the segmented graph are consecutive points in the Cartesian space. The planning method depends on transforming these consecutive points into a graph theory problem, which is a combinatorial optimization problem. The following procedure can explain the methodology:

1. Consider the graph in Fig. 9.12, where each point on the curve has a slope that can be defined as:

$$m = \frac{y_{i+1} - y_i}{x_{i+1} - x_i}, \quad i = 1, \ldots, n, \tag{9.1}$$

where n is the number of points making up the graph and m is the slope of the graph at point i, where i indicates the index of the current point and $i + 1$ indicates the index of the next point on the graph.

2. Using Eq. (9.1), calculate the slope at any point on the graph.
3. Starting from the base point, read the slope and whenever there is a slope change set a node, as shown in Fig. 9.13.

For this problem, the coordinates of the 11 nodes are given to the reader in this step, as shown in Table 9.2.

4. Connect the nodes with each other using as many as possible lines (edges) under the condition that edges should not collide with the obstacles in the field of the work. The result is just a simple directed graph shown in Fig. 9.14. The weights of the edges are their lengths and can be calculated as

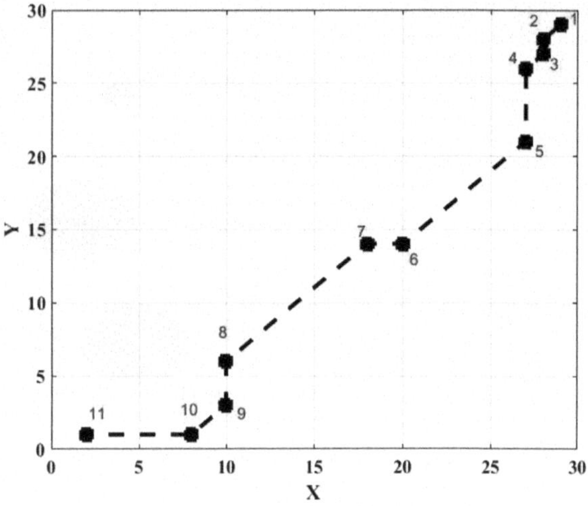

Fig. 9.13 Vertices on the graph represent the points of the slope change

Table 9.2 Coordinates of the ordered vertices

Vertex ID	*x*-axis	*y*-axis	Vertex ID	*x*-axis	*y*-axis
1	29	29	7	18	14
2	28	28	8	10	6
3	28	27	9	10	3
4	27	26	10	8	1
5	27	21	11	2	1
6	20	14			

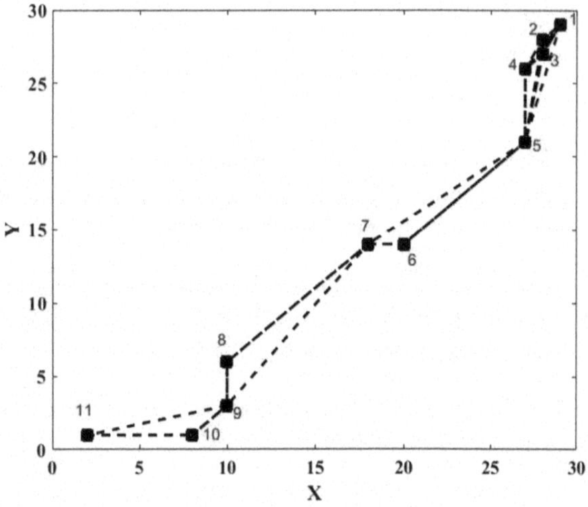

Fig. 9.14 Resultant simple graph after adding possible edges among vertices

$$\text{weight} = \sqrt{(x_2 - x_1)^2 + (y_2 - y_1)^2}. \tag{9.2}$$

5. Find the shortest path on the resulted graph between base point, which is node 1 and task point, which is node 11 in this example. There are many methods to deal with finding the shortest path on a graph, like the Dijkstra algorithm (Dijkstra 1959), Bellman-Ford's algorithm (Cherkassky et al. 1996), the Floyd-Warshall algorithm (Swathika and Hemamalini 2017), or Johnson's algorithm (Pettie 2002). In this book, the graph theory toolbox in MATLAB is proposed as a tool to find the shortest path on a graph. The result obtained from this step is a set of edges with the minimum required lengths connected by minimum required nodes. Each edge corresponds to a link in a robot, and each node corresponds to a revolute joint in a robot.

Since version R2015b, MATLAB has come with properties and functions that are dedicated to graph theory problems. Computing the weights of the edges is the key point for the shortest path algorithm to find the best solution. Table 9.3 shows the IDs of the vertices and the weight of the edges on the graph where vertex 1 ID and vertex 2 ID are the nodes that form an edge.

6. Build the sparse matrix using the *sparse()* function using the following syntax:

$$sparse(vertex\ 1\,column,\ vertex\ 2\,column,\ weight\ column,\ n,\ n),$$

where n is the maximum number of elements in the column vertex 2. Use *graphshortestpath ()* to find the shortest path on the graph by the following syntax:

Table 9.3 Weights of the edges in the directed graph

Edge ID	Vertex 1 ID	Vertex 2 ID	Weight	Edge ID	Vertex 1 ID	Vertex 2 ID	Weight
1	1	2	1.414	11	5	6	9.899
2	1	3	2.236	12	5	7	11.401
3	1	4	3.605	13	6	7	2
4	1	5	8.246	14	7	8	11.313
5	2	3	1	15	7	9	13.601
6	2	4	2.236	16	8	9	3
7	2	5	7.071	17	9	10	2.828
8	3	4	1.414	18	9	11	8.246
9	3	5	6.082	19	10	11	6
10	4	5	5				

$$[dist,\ path] = graphshortestpath(sparse\,matrix,\ starting\,vertex,\ goal\,vertex$$

where *dist* is the total length of the optimum path on the graph. It is worth mentioning that *graphshortestpath* is based on the Dijkstra algorithm and achieves the best walk from vertex to vertex over the given edges. The following code snippets can be useful.

```
DG = sparse(vertex 1 ID,vertex 2 ID,weight,11,11)
h = view(biograph(DG,[],'ShowWeights','on'))
[dist,path,pred] = graphshortestpath(DG,1,11)
set(h.Nodes(path),'Color',[1 0.4 0.4])
edges = getedgesbynodeid(h,get(h.Nodes(path),'ID'));
set(edges,'LineColor',[1 0 0])
set(edges,'LineWidth',1.5)
```

The shortest path on the graph is shown in Figs. 9.15 and 9.16, where it consists of four edges and five nodes. It is worth mentioning that this method works under limitations which are evaluated depending on the designer ingenuity. The workspace should have a relatively low number of constraints, and the coordinates of the base

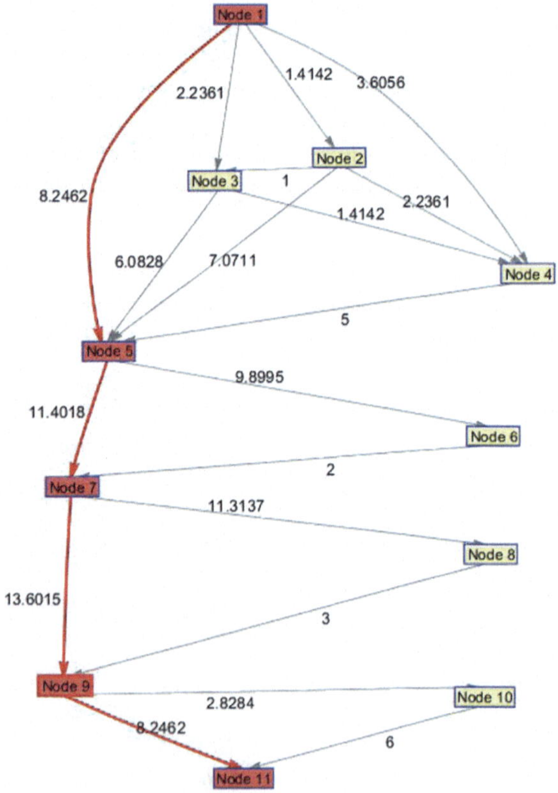

Fig. 9.15 Shortest path on the directed graph

Fig. 9.16 Improved graph versus the original graph

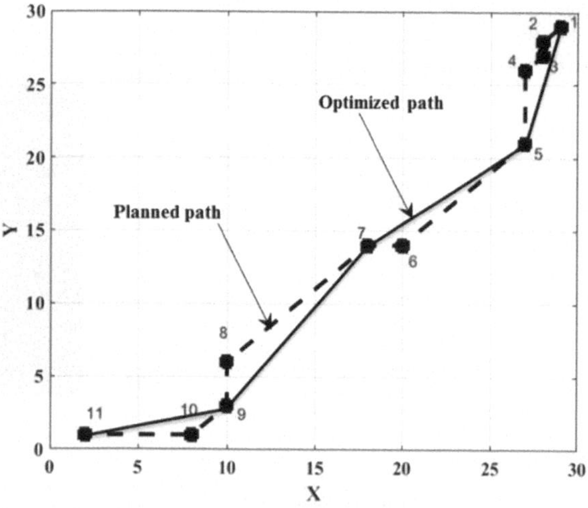

and task points should be selected carefully. The method provides a solution for a single task kinematic synthesis problem. Thus for the multi-task problem, greedy selection may be used to choose the best type that meets all the requirements.

7. By analogy, edges correspond to links, and nodes correspond to joints. As a result, the kinematic synthesis of this example consists of four links with 5DOF, as shown in Fig. 9.17.

Example 9.1 Figure 9.18 shows the requirements that the proposed robot should meet, where there is a single square obstacle and the base point and task point are all defined in the Cartesian space. Applying the procedure of the planning method, the optimum solution is found to be a robot manipulator of 4DOF with three links.

Fig. 9.17 Corresponding robot configuration to the improved heuristic graph

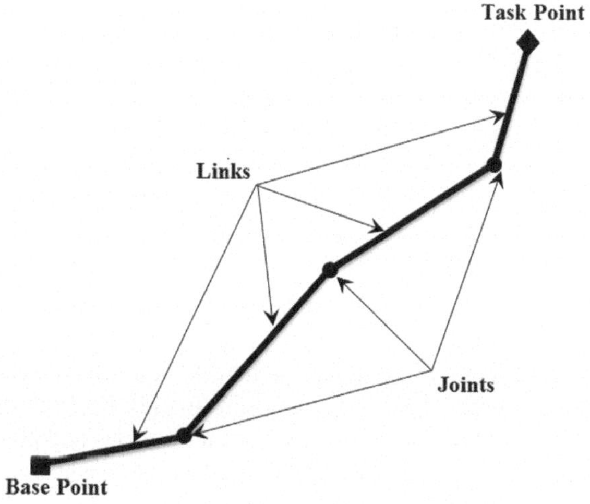

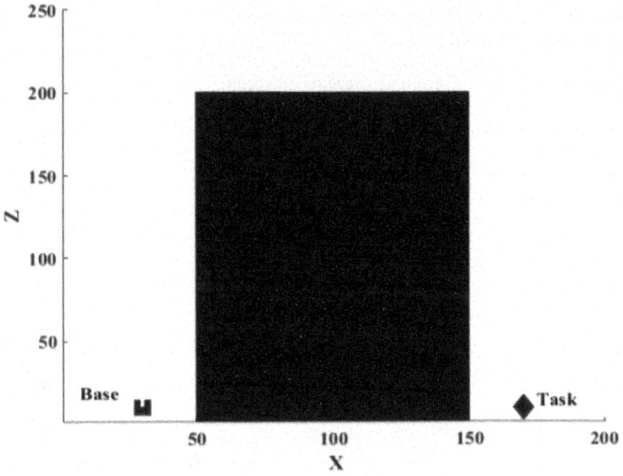

Fig. 9.18 Problem to find a robot that can avoid one obstacle (dark square)

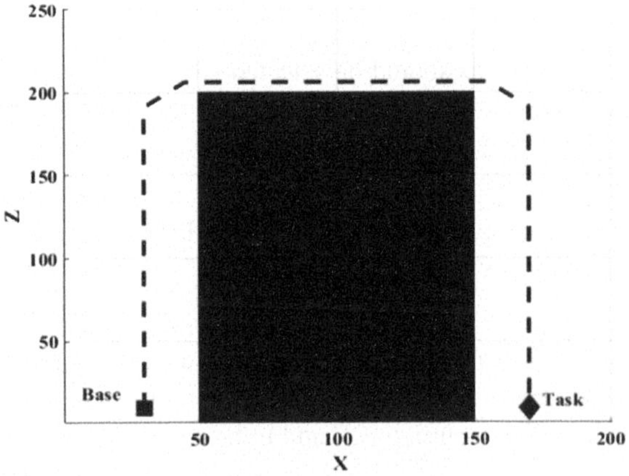

Fig. 9.19 Planned graph between the base and task points in the C-space

Figures 9.19, 9.20, 9.21, and 9.22 explain the intermediate operations to reach the final result. This is a simple application that can easily be carried out by the designer but is a good start to explain the proposed method to find the optimum type synthesis and optimum dimensions for the desired type.

This is a simple application successfully solved automatically by the proposed algorithm without a direct decision from the user or instructor. The search space is:

$$0 \leq x \leq 200$$
$$0 \leq y \leq 250$$

By default, the generated path by A* algorithm should be close to the dark obstacle in Fig. 9.18. There must be a safe margin between the constraint and graph; this margin will be the distance between obstacles and physical links. This can be done easily by including the safe margins in the dimensions of the obstacles themselves.

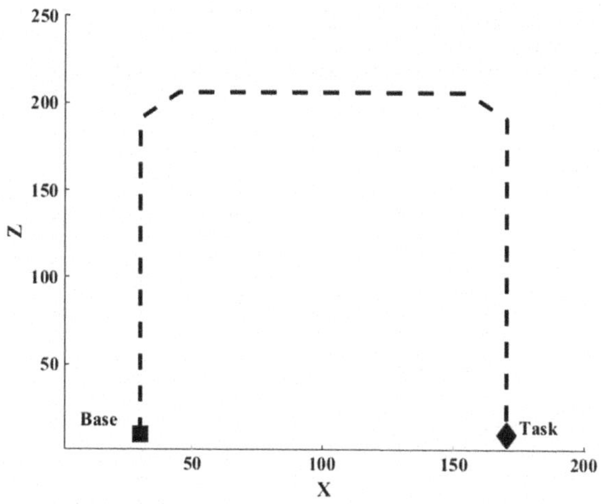

Fig. 9.20 Setting nodes at places of slope directions

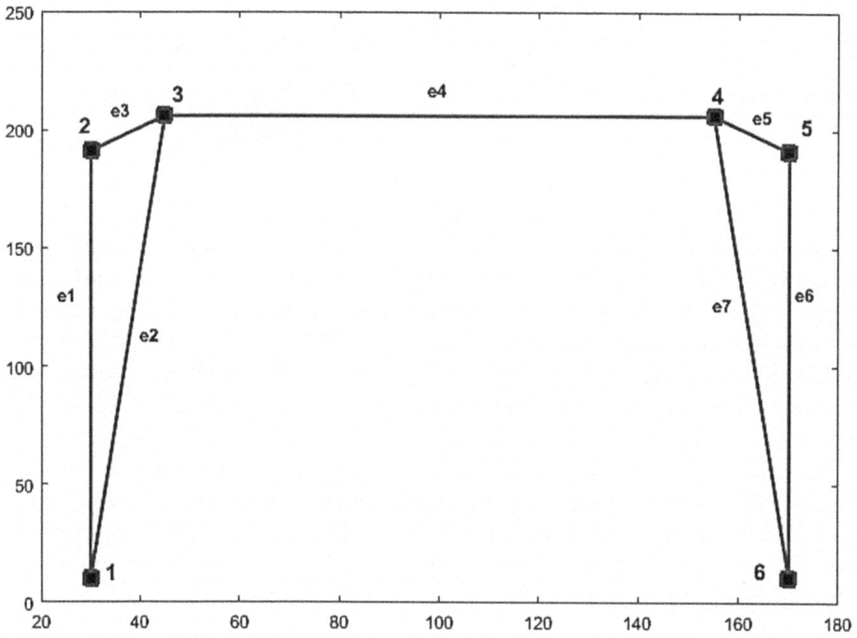

Fig. 9.21 Graph obtained

In this application, the rectangular object dimensions are 90 width +10 margin and 190 height +10 margins. In total, the rectangular object used in this example has 100 × 150 dimensions.

The graph in Fig. 9.20 has six distinguished points that the slope is changed; they should be labelled as nodes. All possible edges connecting the six nodes are shown in

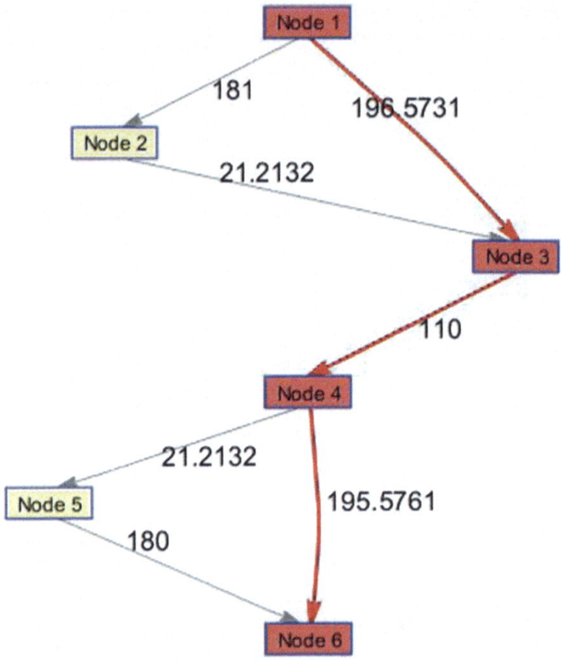

Fig. 9.22 Optimum path on the graph for the given problem

Fig. 9.21 where they are seven edges. The edges are set in a systematic way under the condition that any edge should not touch any obstacle in the workspace; for instance, the software could not set an edge between node 1 and node 4. After calculating the weights of the edges, the software performs a walk on the directed graph from node 1 to node 6 to find the optimum path on the graph. Dijkstra included in graph theory toolbox performs the walk, and the result is shown in Fig. 9.22. As expected, the shortest path on the graph is starting from node 1 to node 3 to node 4 and ends with node 6.

The resulted path has four nodes, and three segments are equivalent to a robot with four joints and three links. The fourth joint can be considered as the wrist of the robot, which does not affect the position of the end-effector but affects its orientation. The simulation model of this problem is revealed in Fig. 9.23, the dimensions of the links are shown in Fig. 9.22.

Example 9.2 Figure 9.24 explains the problem where the base point and task point are given as well as three obstacles, shown in dark colour. It is a more complicated example than the previous one, and the proposed method has given a reasonable solution. Figures 9.25, 9.26, and 9.27 illustrate the operations to the final result, and Fig. 9.28 explains the simulation of the proposed resultant robot. The proposed solution for this problem by the planning method is a robot with five joints and four links. The search space is:

$$0 \leq x \leq 200$$

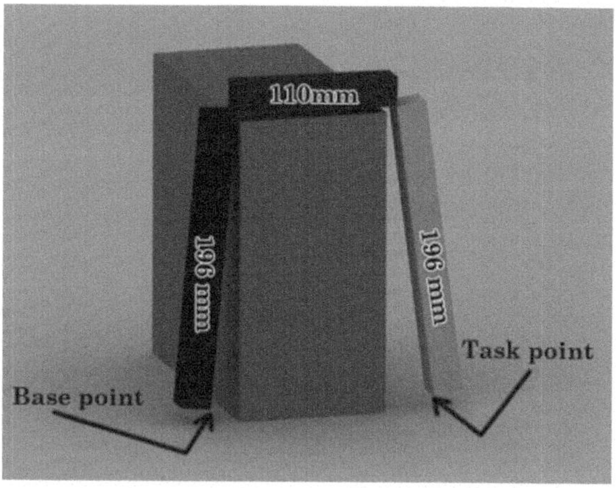

Fig. 9.23 Simulation of the robot kinematic synthesis given by planning method

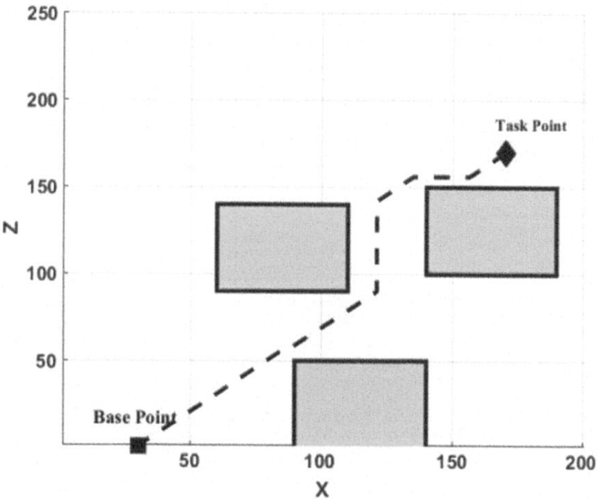

Fig. 9.24 Planned path for Example 9.2

$$0 \leq y \leq 250$$

$$\text{Safe margin} = 10$$

$$\text{Maximum link length} = 65$$

In this example, we have extended the constraints imposed on the search engine to include the maximum length of the link. The open choices of the constraints will be discussed in the next section as tuning parameters and variables that could change the final solution dramatically. For the proposed position of the base point, the graph generated by A* algorithm is shown in Fig. 9.24. The slope of the graph is changed at eight-point, as shown in Fig. 9.25, that could be nodes of the directed graph in Fig. 9.26. Since there is feasible edge set by the software between node 8 and node 5, the resultant shortest path on the graph will as shown in Fig. 9.27 dropping node 6

Fig. 9.25 Path with nodes

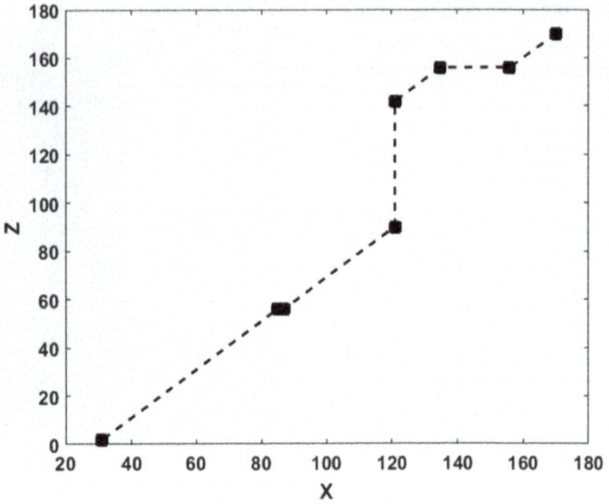

Fig. 9.26 Possible segments between nodes in the path of Example 9.2

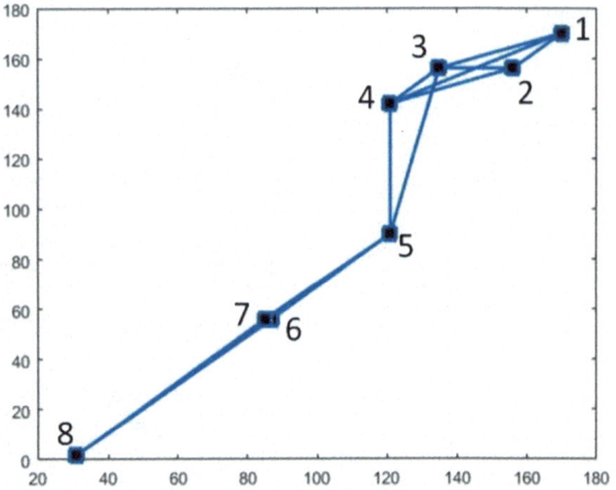

and node 7. This will lead to a segment of length 125.873 between node 5 and node 8. Since length of 125.873 is a violation of the constraints, the algorithm breaks this segment into two equal segments each one has 62.9365 adding extra node between nodes 5 and 8. The resultant graph has five nodes and four segments that correspond to a robot with five joints and four segments. The simulation model is shown in Fig. 9.28 with links lengths shown in Fig. 9.27.

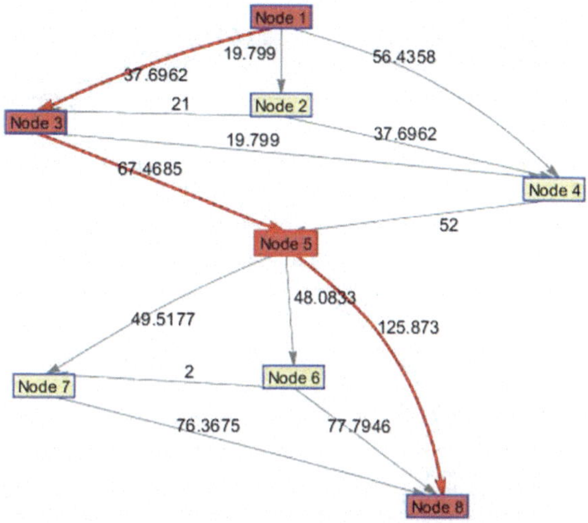

Fig. 9.27 Minimum segments needed to reach the task point

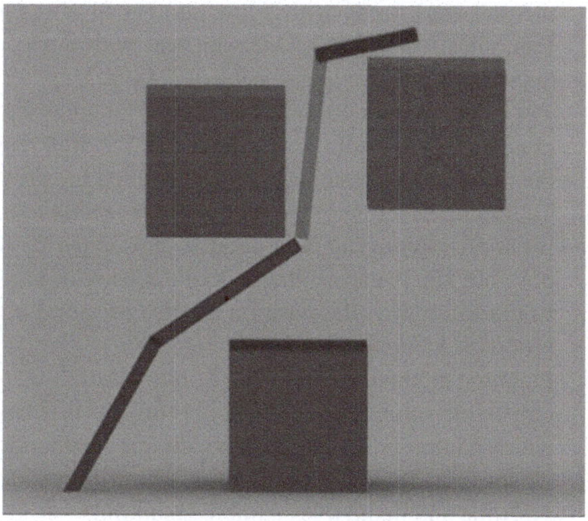

Fig. 9.28 Simulation for the resulting robot for Example 9.2

9.9 Discussion on Planning Method

The proposed method to find the kinematic synthesis of a serial robot manipulator using the general procedure mentioned in Sect. 9.8 can be tuned and restricted by the following parameters and variables

1. The safe margin is a tuning parameter and a distance between physical link and obstacle in the workspace, and it varies from problem to another. Safe margin can be added to the dimensions of the obstacles, in case of circular objects it summed

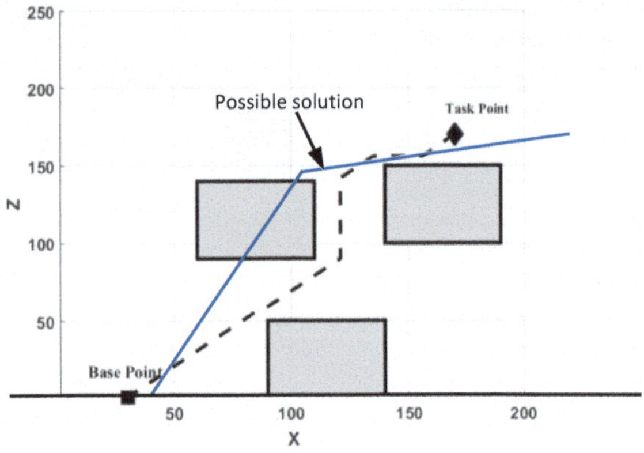

Fig. 9.29 Effect of the tuning of the position of the base point on the final kinematic synthesis

to the radius of the object, and in case of a rectangular object, it summed to the width and length.

2. The number of obstacles is a design variable set by the user or instructor. This method works only with a low number of obstacle objects as we have seen in Examples 9.1 and 9.2. One can start with a specific number of obstacles and decrease it until finding feasible solution.
3. The maximum length of a link can be tuned as we have seen in Example 9.2 to obtain a feasible solution. In Example 9.2, link length of 125.873 will not allow the proposed kinematic to work and reach the task point.
4. Tuning the position of the base point is crucial to the final solution, for instance, if the position of the base point is chosen (−15) for Example 9.2 problem, the possible kinematic synthesis could consist of two links as shown in Fig. 9.29. In the same way, the position of the task point affects the final solution of the proposed method.
5. Planning method may return the optimal solution, nonoptimal solution, and even unfeasible solution for a specific problem. It is still a useful automated method which can propose a candidate solution. Furthermore, the results obtained from planning method can be used as an initial solution for Liu method to reduce run-time and make it less time-consuming.
6. Planning method works in 2D and 3D space to find the kinematic synthesis of robots, in contrast Liu approach, works only for planar mechanism.

Tuning the parameters of an algorithm is a usual process for all optimization algorithms, and one can tune the population size or number of iteration or any other parameters to find the best solution for a specific optimization problem. Planning method is a hybrid system that consists of A* algorithm, graph theory toolbox and sequential software aimed to find an automatic method to find the kinematic synthesis of a robot.

9.10 Summary

Kinematic synthesis was discussed in this chapter where the problem is to find an appropriate robot type with minimum dimensionality for a specific task in a systematic and automated way. This problem was done usually by the designer; however, this chapter introduces the evolutionary method and planning method to deal with this problem using computer science with less depending on the human being.

References

Bollobás B (2013) Modern graph theory, vol 184. Springer Science & Business Media, Berlin

Cherkassky BV, Goldberg AV, Radzik T (1996) Shortest paths algorithms: theory and experimental evaluation. Math Program 73(2):129–174

Dijkstra EW (1959) A note on two problems in connexion with graphs. Numer Math 1(1):269–271

Liu Y, McPhee J (2007) Automated kinematic synthesis of planar mechanisms with revolute joints. Mech Based Des Struct Mach 35(4):405–445

Mühlenbein H (1997) Genetic algorithms. In: Aarts E, Lenstra JK (eds) Local search in combinatorial optimization. John Wiley & Sons, New York, pp 137–172

Oliva JC, Goodman ED (2010) Simultaneous type and dimensional synthesis of Planar 1DOF mechanisms using evolutionary search and convertible agents (DETC2009-86722). J Mech Robot 2(3):031001

Pettie S (2002) A faster all-pairs shortest path algorithm for real-weighted sparse graphs. In: International colloquium on automata, languages, and programming. Springer, Berlin, pp 85–97

Swathika OG, Hemamalini S (2017) Prims aided Floyd Warshall algorithm for shortest path identification in microgrid. In: Emerging trends in electrical, communications and information technologies. Springer, Berlin, pp 283–291

Thiele R (2005) The mathematics and science of Leonhard Euler (1707–1783). In: Mathematics and the Historian's Craft. Springer, Berlin, pp 81–140

West DB (2001) Introduction to graph theory, vol 2. Prentice Hall, Upper Saddle River

Index

H. N. Ghafil and K. Jármai (ed.), *Optimization for Robot Modelling with MATLAB*,
https://doi.org/10.1007/978-3-030-40410-9

The manufacturer's authorised representative in the EU is Springer Nature Customer Service Centre GmbH, Europaplatz 3, 69115 Heidelberg, Germany. If you have any concerns regarding our products, please contact ProductSafety@springernature.com

Printed and bound by CPI Group (UK) Ltd, Croydon, CR0 4YY
30/09/2026
02225449-0008